ISEE Upper-Level

Subject Test Mathematics

Student Practice Workbook

+ Two Full-Length ISEE Upper-Level Math Tests

SCAN ME

Math Notion

www.MathNotion.com

ISEE Upper-Level Subject Test Mathematics

Published in the United State of America By

The Math Notion

Web: WWW.MathNotion.com

Email: info@Mathnotion.com

ISBN: 978-1-63620-060-6

The Math Notion

Michael Smith has been a math instructor for over a decade now. He launched the Math Notion. Since 2006, we have devoted our time to both teaching and developing exceptional math learning materials. As a test prep company, we have worked with thousands of students. We have used the feedback of our students to develop a unique study program that can be used by students to drastically improve their math scores fast and effectively. We have more than a thousand Math learning books including:

– **SAT Math Prep**

– **ACT Math Prep**

– **SSAT/ISEE Math Prep**

– **Accuplacer Math Prep**

– **Common Core Math Prep**

–**many Math Education Workbooks, Study Guides, Practice and Exercise Books**

As an experienced Math test preparation company, we have helped many students raise their standardized test scores—and attend the colleges of their dreams: We tutor online and in person, we teach students in large groups, and we provide training materials and textbooks through our website and through Amazon.

You can contact us via email at:

info@Mathnotion.com

Get the Targeted Practice You Need to Ace the ISEE Upper-Level Math Test!

ISEE Upper-Level Subject Test Mathematics includes easy-to-follow instructions, helpful examples, and plenty of math practice problems to assist students to master each concept, brush up their problem-solving skills, and create confidence.

The ISEE Upper-Level math practice book provides numerous opportunities to evaluate basic skills along with abundant remediation and intervention activities. It is a skill that permits you to quickly master intricate information and produce better leads in less time.

Students can boost their test-taking skills by taking the book's two practice ISEE Upper-Level Math exams. All test questions answered and explained in detail.

Important Features of the ISEE Upper-Level Math Book:

- A **complete review** of ISEE Upper-Level math test topics,
- Over 2,500 practice problems covering all topics tested,
- The most important concepts you need to know,
- Clear and concise, easy-to-follow sections,
- Well designed for enhanced learning and interest,
- Hands-on experience with all question types,
- **2 full-length practice tests** with detailed answer explanations,
- Cost-Effective Pricing,

Powerful math exercises to help you avoid traps and pacing yourself to beat the ISEE Upper-Level test. Students will gain valuable experience and raise their confidence by taking math practice tests, learning about test structure, and gaining a deeper understanding of what is tested on the ISEE Upper-Level Math. If ever there was a book to respond to the pressure to increase students' test scores, this is it.

WWW.MathNotion.COM

… So Much More Online!

✓ FREE Math Lessons

✓ More Math Learning Books!

✓ Mathematics Worksheets

✓ Online Math Tutors

For a PDF Version of This Book

SCAN ME

Please Visit WWW.MathNotion.com

Contents

Chapter 1 :
Integers and Number Theory

Topics that you will practice in this chapter:

- ✓ Rounding
- ✓ Whole Number Addition and Subtraction
- ✓ Whole Number Multiplication and Division
- ✓ Rounding and Estimates
- ✓ Adding and Subtracting Integers
- ✓ Multiplying and Dividing Integers
- ✓ Order of Operations
- ✓ Ordering Integers and Numbers
- ✓ Integers and Absolute Value
- ✓ Factoring Numbers
- ✓ Greatest Common Factor (GCF)
- ✓ Least Common Multiple (LCM)

"Wherever there is number, there is beauty." –*Proclus*

Rounding

✎ **Round each number to the nearest ten.**

1) 42 = ____   5) 19 = ____   9) 48 = ____

2) 88 = ____   6) 25 = ____   10) 81 = ____

3) 24 = ____   7) 93 = ____   11) 58 = ____

4) 57 = ____   8) 71 = ____   12) 87 = ____

✎ **Round each number to the nearest hundred.**

13) 198 = ____   17) 321 = ____   21) 580 = ____

14) 387 = ____   18) 433 = ____   22) 868 = ____

15) 816 = ____   19) 579 = ____   23) 480 = ____

16) 101 = ____   20) 825 = ____   24) 287 = ____

✎ **Round each number to the nearest thousand.**

25) 1,382 = ____   29) 9,099 = ____   33) 52,866 = ____

26) 3,420 = ____   30) 22,980 = ____   34) 85,190 = ____

27) 4,254 = ____   31) 45,188 = ____   35) 70,990 = ____

28) 6,861 = ____   32) 16,808 = ____   36) 26,869 = ____

Rounding and Estimates

✍ **Estimate the sum by rounding each number to the nearest ten.**

1) $13 + 22 =$ _____

2) $71 + 23 =$ _____

3) $61 + 58 =$ _____

4) $56 + 85 =$ _____

5) $368 + 249 =$ _____

6) $330 + 903 =$ _____

7) $471 + 293 =$ _____

8) $1,950 + 2,655 =$ _____

✍ **Estimate the product by rounding each number to the nearest ten.**

9) $32 \times 71 =$ _____

10) $12 \times 33 =$ _____

11) $31 \times 83 =$ _____

12) $19 \times 11 =$ _____

13) $42 \times 76 =$ _____

14) $63 \times 34 =$ _____

15) $19 \times 31 =$ _____

16) $59 \times 71 =$ _____

✍ **Estimate the sum or product by rounding each number to the nearest ten.**

17) $\begin{array}{r} 29 \\ \times\ 12 \\ \hline \\ \hline \end{array}$

18) $\begin{array}{r} 37 \\ \times\ 26 \\ \hline \\ \hline \end{array}$

19) $\begin{array}{r} 48 \\ +\ 82 \\ \hline \\ \hline \end{array}$

20) $\begin{array}{r} 65 \\ +44 \\ \hline \\ \hline \end{array}$

21) $\begin{array}{r} 37 \\ \times\ 14 \\ \hline \\ \hline \end{array}$

22) $\begin{array}{r} 71 \\ +\ 32 \\ \hline \\ \hline \end{array}$

Adding and Subtracting Integers

✎ **Find each sum.**

1) $14 + (-6) =$

2) $(-13) + (-20) =$

3) $5 + (-28) =$

4) $50 + (-12) =$

5) $(-7) + (-15) + 3 =$

6) $30 + (-14) + 8 =$

7) $40 + (-10) + (-14) + 17 =$

8) $(-15) + (-20) + 13 + 35 =$

9) $40 + (-20) + (38 - 29) =$

10) $28 + (-12) + (30 - 12) =$

✎ **Find each difference.**

11) $(-18) - (-7) =$

12) $25 - (-14) =$

13) $(-20) - 36 =$

14) $34 - (-19) =$

15) $51 - (30 - 21) =$

16) $17 - (5) - (-24) =$

17) $(35 + 20) - (-46) =$

18) $48 - 16 - (-8) =$

19) $62 - (28 + 17) - (-15) =$

20) $58 - (-23) - (-31) =$

21) $19 - (-8) - (-13) =$

22) $(19 - 24) - (-14) =$

23) $27 - 33 - (-21) =$

24) $58 - (32 + 24) - (-9) =$

25) $36 - (-30) + (-17) =$

26) $27 - (-42) + (-31) =$

Multiplying and Dividing Integers

✎ **Find each product.**

1) $(-9) \times (-5) =$

2) $(-3) \times 9 =$

3) $8 \times (-12) =$

4) $(-7) \times (-20) =$

5) $(-3) \times (-5) \times 6 =$

6) $(14 - 3) \times (-8) =$

7) $12 \times (-9) \times (-3) =$

8) $(140 + 10) \times (-2) =$

9) $10 \times (-12 + 8) \times 3 =$

10) $(-8) \times (-5) \times (-10) =$

✎ **Find each quotient.**

11) $42 \div (-7) =$

12) $(-48) \div (-6) =$

13) $(-40) \div (-8) =$

14) $54 \div (-2) =$

15) $152 \div 19 =$

16) $(-144) \div (-12) =$

17) $180 \div (-10) =$

18) $(-312) \div (-12) =$

19) $221 \div (-13) =$

20) $(-126) \div (6) =$

21) $(-161) \div (-7) =$

22) $-266 \div (-14) =$

23) $(-120) \div (-4) =$

24) $270 \div (-18) =$

25) $(-208) \div (-8) =$

26) $(135) \div (-15) =$

Order of Operations

✎ **Evaluate each expression.**

1) $7 + (5 \times 4) =$

2) $14 - (3 \times 6) =$

3) $(19 \times 4) + 16 =$

4) $(16 - 7) - (8 \times 2) =$

5) $27 + (18 \div 3) =$

6) $(18 \times 8) \div 6 =$

7) $(32 \div 4) \times (-2) =$

8) $(9 \times 4) + (32 - 18) =$

9) $24 + (4 \times 3) + 7 =$

10) $(36 \times 3) \div (2 + 2) =$

11) $(-7) + (12 \times 3) + 11 =$

12) $(8 \times 5) - (24 \div 6) =$

13) $(7 \times 6 \div 3) - (12 + 9) =$

14) $(13 + 5 - 14) \times 3 - 2 =$

15) $(20 - 14 + 30) \times (64 \div 4) =$

16) $32 + \big(28 - (36 \div 9)\big) =$

17) $(7 + 6 - 4 - 7) + (15 \div 5) =$

18) $(85 - 20) + (20 - 18 + 7) =$

19) $(20 \times 2) + (14 \times 3) - 22 =$

20) $18 + 5 - (30 \times 3) + 20 =$

Ordering Integers and Numbers

✎ **Order each set of integers from least to greatest.**

1) $8, -10, -5, -3, 4$ ___, ___, ___, ___, ___, ___

2) $-10, -18, 6, 14, 27$ ___, ___, ___, ___, ___, ___

3) $15, -8, -21, 21, -23$ ___, ___, ___, ___, ___, ___

4) $-14, -40, 23, -12, 47$ ___, ___, ___, ___, ___, ___

5) $59, -54, 32, -57, 36$ ___, ___, ___, ___, ___, ___

6) $68, 26, -19, 47, -34$ ___, ___, ___, ___, ___, ___

✎ **Order each set of integers from greatest to least.**

7) $18, 36, -16, -18, -10$ ___, ___, ___, ___, ___, ___

8) $27, 34, -12, -24, 94$ ___, ___, ___, ___, ___, ___

9) $50, -21, -13, 42, -2$ ___, ___, ___, ___, ___, ___

10) $37, 46, -20, -16, 86$ ___, ___, ___, ___, ___, ___

11) $-18, 88, -26, -59, 75$ ___, ___, ___, ___, ___, ___

12) $-65, -30, -25, 3, 14$ ___, ___, ___, ___, ___, ___

Integers and Absolute Value

✍ **Write absolute value of each number.**

1) $|-2| =$

2) $|-27| =$

3) $|-20| =$

4) $|14| =$

5) $|6| =$

6) $|-55| =$

7) $|16| =$

8) $|2| =$

9) $|54| =$

10) $|-4| =$

11) $|-11|$

12) $|88| =$

13) $|0| =$

14) $|79| =$

15) $|-32| =$

16) $|-17| =$

17) $|42| =$

18) $|-46| =$

19) $|1| =$

20) $|-40| =$

✍ **Evaluate the value.**

21) $|-5| - \dfrac{|-21|}{7} =$

22) $14 - |3 - 15| - |-4| =$

23) $\dfrac{|-32|}{4} \times |-4| =$

24) $\dfrac{|7 \times (-3)|}{7} \times \dfrac{|-19|}{3} =$

25) $|4 \times (-5)| + \dfrac{|-40|}{5} =$

26) $\dfrac{|-45|}{9} \times \dfrac{|-24|}{12} =$

27) $|-12 + 8| \times \dfrac{|-7 \times 7|}{7}$

28) $\dfrac{|-11 \times 2|}{4} \times |-16| =$

Factoring Numbers

✍ **List all positive factors of each number.**

1) 9	9) 28	17) 50
2) 16	10) 98	18) 62
3) 24	11) 14	19) 95
4) 30	12) 54	20) 64
5) 26	13) 55	21) 70
6) 46	14) 18	22) 45
7) 20	15) 63	23) 22
8) 68	16) 34	24) 65

Greatest Common Factor

✎ **Find the GCF for each number pair.**

1) 6, 2

2) 4, 5

3) 3, 12

4) 7, 3

5) 5, 10

6) 8, 48

7) 6, 18

8) 9, 15

9) 12, 18

10) 4, 36

11) 6, 10

12) 28, 52

13) 25, 10

14) 22, 24

15) 9, 54

16) 8, 54

17) 42, 14

18) 16, 40

19) 9, 2, 3

20) 5, 15, 10

21) 7, 9, 2

22) 16, 64

23) 30, 48

24) 36, 63

Least Common Multiple

✎ **Find the LCM for each number pair.**

1) 6, 9

2) 15, 45

3) 16, 40

4) 12, 36

5) 18, 27

6) 14, 42

7) 6, 30

8) 8, 56

9) 7, 21

10) 8, 20

11) 15, 25

12) 7, 9

13) 4, 11

14) 8, 28

15) 28, 56

16) 40, 50

17) 12, 13

18) 22, 11

19) 36, 20

20) 15, 35

21) 18, 81

22) 30, 54

23) 18, 45

24) 75, 25

Answers of Worksheets

Rounding

1) 40	10) 80	19) 600	28) 7,000
2) 90	11) 60	20) 800	29) 9,000
3) 20	12) 90	21) 600	30) 23,000
4) 60	13) 200	22) 900	31) 45,000
5) 20	14) 400	23) 500	32) 17,000
6) 30	15) 800	24) 300	33) 53,000
7) 90	16) 100	25) 1,000	34) 85,000
8) 70	17) 300	26) 3,000	35) 71,000
9) 50	18) 400	27) 4,000	36) 27,000

Rounding and Estimates

1) 30	7) 760	13) 3,200	19) 130
2) 90	8) 4,610	14) 1,800	20) 110
3) 120	9) 2,100	15) 600	21) 400
4) 150	10) 300	16) 4,200	22) 100
5) 620	11) 2,400	17) 300	
6) 1,230	12) 200	18) 1,200	

Adding and Subtracting Integers

1) 8	8) 13	15) 42	22) 9
2) −33	9) 29	16) 36	23) 15
3) −23	10) 34	17) 101	24) 11
4) 38	11) −11	18) 40	25) 49
5) −19	12) 39	19) 32	26) 38
6) 24	13) −56	20) 112	
7) 33	14) 53	21) 40	

Multiplying and Dividing Integers

1) 45	6) −88	11) −6	16) 12
2) −27	7) 324	12) 8	17) −18
3) −96	8) −300	13) 5	18) 26
4) 140	9) −120	14) −27	19) −17
5) 90	10) −400	15) 8	20) −21

21) 23	23) 30	25) 26
22) 19	24) -15	26) -9

Order of Operations

1) 27	6) 24	11) 40	16) 56
2) -4	7) -16	12) 36	17) 5
3) 92	8) 50	13) -7	18) 74
4) -7	9) 43	14) 10	19) 60
5) 33	10) 27	15) 576	20) -47

Ordering Integers and Numbers

1) $-10, -5, -3, 4, 8$	7) $36, 18, -10, -16, -18$
2) $-18, -10, 6, 14, 27$	8) $94, 34, 27, -12, -24$
3) $-23, -21, -8, 15, 21$	9) $50, 42, -2, -13, -21$
4) $-40, -14, -12, 23, 47$	10) $86, 46, 37, -16, -20$
5) $-57, -54, 32, 36, 59$	11) $88, 75, -18, -26, -59$
6) $-34, -19, 26, 47, 68$	12) $14, 3, -25, -30, -65$

Integers and Absolute Value

1) 2	8) 2	15) 32	22) -2
2) 27	9) 54	16) 17	23) 32
3) 20	10) 4	17) 42	24) 19
4) 14	11) 11	18) 46	25) 28
5) 6	12) 88	19) 1	26) 10
6) 55	13) 0	20) 40	27) 28
7) 16	14) 79	21) 2	28) 88

Factoring Numbers

1) $1, 3, 9$	9) $1, 2, 4, 7, 14, 28$	17) $1, 2, 5, 10, 25, 50$
2) $1, 2, 4, 8, 16$	10) $1, 2, 7, 14, 49, 98$	18) $1, 2, 31, 62$
3) $1, 2, 3, 4, 6, 8, 12, 24$	11) $1, 2, 7, 14$	19) $1, 5, 19, 95$
4) $1, 2, 3, 5, 6, 10, 15, 30$	12) $1, 2, 3, 6, 9, 18, 27, 54$	20) $1, 2, 4, 8, 16, 32, 64$
5) $1, 2, 13, 26$	13) $1, 5, 11, 55$	21) $1, 2, 5, 7, 10, 14, 35, 70$
6) $1, 2, 23, 46$	14) $1, 2, 3, 6, 9, 18$	22) $1, 3, 5, 9, 15, 45$
7) $1, 2, 4, 5, 10, 20$	15) $1, 3, 7, 9, 21, 63$	23) $1, 2, 11, 22$
8) $1, 2, 4, 17, 34, 68$	16) $1, 2, 17, 34$	24) $1, 5, 13, 65$

Greatest Common Factor

1) 2	7) 6	13) 5	19) 1
2) 1	8) 3	14) 2	20) 5
3) 3	9) 6	15) 9	21) 1
4) 1	10) 4	16) 2	22) 16
5) 5	11) 2	17) 14	23) 6
6) 8	12) 4	18) 8	24) 9

Least Common Multiple

1) 18	7) 30	13) 44	19) 180
2) 45	8) 56	14) 56	20) 105
3) 80	9) 21	15) 56	21) 162
4) 36	10) 40	16) 200	22) 270
5) 54	11) 75	17) 156	23) 90
6) 42	12) 63	18) 22	24) 75

Chapter 2 :

Fractions and Decimals

Topics that you will practice in this chapter:

- ✓ Simplifying Fractions
- ✓ Adding and Subtracting Fractions
- ✓ Multiplying and Dividing Fractions
- ✓ Adding and Subtract Mixed Numbers
- ✓ Multiplying and Dividing Mixed Numbers
- ✓ Adding and Subtracting Decimals
- ✓ Multiplying and Dividing Decimals
- ✓ Comparing Decimals
- ✓ Rounding Decimals

"A Man is like a fraction whose numerator is what he is and whose denominator is what he thinks of himself. The larger the denominator, the smaller the fraction." –Tolstoy

Simplifying Fractions

✎ **Simplify each fraction to its lowest terms.**

1) $\frac{5}{10} =$

2) $\frac{28}{35} =$

3) $\frac{27}{36} =$

4) $\frac{40}{80} =$

5) $\frac{14}{56} =$

6) $\frac{32}{48} =$

7) $\frac{52}{65} =$

8) $\frac{15}{60} =$

9) $\frac{80}{160} =$

10) $\frac{55}{77} =$

11) $\frac{28}{112} =$

12) $\frac{32}{64} =$

13) $\frac{63}{72} =$

14) $\frac{81}{90} =$

15) $\frac{35}{105} =$

16) $\frac{25}{70} =$

17) $\frac{80}{280} =$

18) $\frac{12}{81} =$

19) $\frac{36}{186} =$

20) $\frac{240}{540} =$

21) $\frac{70}{560} =$

✎ **Find the answer for each problem.**

22) Which of the following fractions equal to $\frac{3}{4}$? _____

     A. $\frac{60}{90}$
     B. $\frac{43}{104}$
     C. $\frac{48}{64}$
     D. $\frac{150}{300}$

23) Which of the following fractions equal to $\frac{5}{8}$? _____

     A. $\frac{125}{200}$
     B. $\frac{115}{200}$
     C. $\frac{50}{100}$
     D. $\frac{30}{90}$

24) Which of the following fractions equal to $\frac{3}{7}$? _____

     A. $\frac{58}{116}$
     B. $\frac{54}{126}$
     C. $\frac{270}{167}$
     D. $\frac{42}{63}$

Adding and Subtracting Fractions

✎ **Find the sum.**

1) $\frac{5}{9} + \frac{4}{9} =$

2) $\frac{1}{2} + \frac{1}{7} =$

3) $\frac{3}{8} + \frac{1}{4} =$

4) $\frac{3}{5} + \frac{1}{2} =$

5) $\frac{1}{4} + \frac{3}{5} =$

6) $\frac{7}{8} + \frac{3}{8} =$

7) $\frac{1}{2} + \frac{7}{10} =$

8) $\frac{2}{5} + \frac{2}{3} =$

9) $\frac{5}{7} + \frac{2}{3} =$

10) $\frac{7}{12} + \frac{3}{4} =$

11) $\frac{5}{6} + \frac{2}{5} =$

12) $\frac{1}{12} + \frac{2}{3} =$

✎ **Find the difference.**

13) $\frac{1}{3} - \frac{1}{6} =$

14) $\frac{3}{4} - \frac{1}{8} =$

15) $\frac{1}{2} - \frac{1}{3} =$

16) $\frac{1}{4} - \frac{1}{5} =$

17) $\frac{5}{8} - \frac{2}{3} =$

18) $\frac{1}{4} - \frac{1}{7} =$

19) $\frac{5}{6} - \frac{1}{9} =$

20) $\frac{3}{4} - \frac{1}{6} =$

21) $\frac{7}{8} - \frac{1}{12} =$

22) $\frac{8}{15} - \frac{3}{5} =$

23) $\frac{3}{12} - \frac{1}{14} =$

24) $\frac{10}{13} - \frac{7}{26} =$

25) $\frac{6}{7} - \frac{3}{4} =$

26) $\frac{4}{5} - \frac{1}{8} =$

27) $\frac{4}{7} - \frac{2}{35} =$

28) $\frac{9}{16} - \frac{2}{8} =$

29) $\frac{8}{9} - \frac{7}{18} =$

30) $\frac{1}{2} - \frac{4}{9} =$

Multiplying and Dividing Fractions

✎ **Find the value of each expression in lowest terms.**

1) $\frac{1}{5} \times \frac{15}{5} =$

2) $\frac{9}{12} \times \frac{4}{9} =$

3) $\frac{1}{16} \times \frac{8}{10} =$

4) $\frac{1}{24} \times \frac{8}{10} =$

5) $\frac{1}{5} \times \frac{1}{4} =$

6) $\frac{7}{9} \times \frac{1}{7} =$

7) $\frac{6}{7} \times \frac{1}{3} =$

8) $\frac{2}{8} \times \frac{2}{8} =$

9) $\frac{5}{8} \times \frac{3}{5} =$

10) $\frac{4}{7} \times \frac{1}{8} =$

11) $\frac{7}{15} \times \frac{5}{7} =$

12) $\frac{3}{10} \times \frac{5}{9} =$

✎ **Find the value of each expression in lowest terms.**

13) $\frac{1}{4} \div \frac{1}{8} =$

14) $\frac{1}{10} \div \frac{1}{5} =$

15) $\frac{3}{4} \div \frac{1}{5} =$

16) $\frac{1}{3} \div \frac{5}{6} =$

17) $\frac{1}{7} \div \frac{8}{42} =$

18) $\frac{3}{4} \div \frac{1}{6} =$

19) $\frac{2}{7} \div \frac{7}{13} =$

20) $\frac{1}{24} \div \frac{3}{16} =$

21) $\frac{7}{12} \div \frac{5}{6} =$

22) $\frac{22}{18} \div \frac{11}{9} =$

23) $\frac{9}{35} \div \frac{3}{7} =$

24) $\frac{2}{7} \div \frac{8}{21} =$

25) $\frac{1}{9} \div \frac{2}{5} =$

26) $\frac{5}{12} \div \frac{3}{5} =$

27) $\frac{3}{20} \div \frac{1}{6} =$

28) $\frac{8}{20} \div \frac{3}{4} =$

29) $\frac{5}{6} \div \frac{2}{9} =$

30) $\frac{5}{11} \div \frac{3}{4} =$

Adding and Subtracting Mixed Numbers

✏ **Find the sum.**

1) $3\frac{1}{3} + 2\frac{1}{6} =$

2) $4\frac{1}{2} + 3\frac{1}{2} =$

3) $3\frac{3}{8} + 1\frac{1}{8} =$

4) $2\frac{1}{4} + 2\frac{1}{3} =$

5) $3\frac{5}{6} + 2\frac{7}{12} =$

6) $5\frac{4}{15} + 3\frac{3}{5} =$

7) $2\frac{1}{3} + 4\frac{3}{7} =$

8) $3\frac{1}{2} + 4\frac{2}{5} =$

9) $5\frac{2}{5} + 6\frac{3}{7} =$

10) $8\frac{5}{16} + 6\frac{1}{12} =$

✏ **Find the difference.**

11) $3\frac{1}{4} - 1\frac{3}{4} =$

12) $6\frac{3}{5} - 4\frac{2}{5} =$

13) $4\frac{1}{3} - 3\frac{1}{9} =$

14) $7\frac{1}{7} - 5\frac{1}{2} =$

15) $5\frac{1}{3} - 2\frac{1}{12} =$

16) $8\frac{1}{5} - 4\frac{1}{3} =$

17) $9\frac{1}{4} - 6\frac{1}{8} =$

18) $11\frac{7}{15} - 8\frac{3}{5} =$

19) $14\frac{5}{6} - 11\frac{3}{5} =$

20) $18\frac{2}{7} - 14\frac{1}{5} =$

21) $9\frac{1}{3} - 4\frac{1}{4} =$

22) $6\frac{1}{8} - 4\frac{1}{16} =$

23) $19\frac{3}{8} - 15\frac{1}{3} =$

24) $11\frac{1}{9} - 8\frac{1}{8} =$

25) $17\frac{1}{7} - 11\frac{1}{5} =$

26) $16\frac{2}{9} - 9\frac{5}{7} =$

Multiplying and Dividing Mixed Numbers

✎ **Find the product.**

1) $5\frac{1}{2} \times 2\frac{1}{4} =$

2) $5\frac{1}{3} \times 4\frac{1}{3} =$

3) $5\frac{3}{4} \times 6\frac{1}{4} =$

4) $3\frac{1}{3} \times 2\frac{3}{5} =$

5) $4\frac{8}{10} \times 1\frac{1}{24} =$

6) $6\frac{2}{7} \times 1\frac{1}{11} =$

7) $8\frac{2}{3} \times 3\frac{1}{2} =$

8) $3\frac{4}{7} \times 2\frac{1}{5} =$

9) $5\frac{2}{8} \times 4\frac{1}{6} =$

10) $7\frac{3}{3} \times 1\frac{3}{8} =$

✎ **Find the quotient.**

11) $2\frac{2}{5} \div 4\frac{1}{5} =$

12) $4\frac{1}{6} \div 3\frac{1}{3} =$

13) $6\frac{1}{3} \div 1\frac{1}{2} =$

14) $7\frac{1}{10} \div 2\frac{2}{5} =$

15) $3\frac{1}{3} \div 1\frac{1}{9} =$

16) $1\frac{1}{10} \div 4\frac{1}{2} =$

17) $1\frac{3}{16} \div 5\frac{1}{4} =$

18) $4\frac{1}{3} \div 4\frac{3}{4} =$

19) $9\frac{1}{3} \div 2\frac{1}{4} =$

20) $15\frac{1}{3} \div 5\frac{1}{2} =$

21) $4\frac{1}{6} \div 1\frac{1}{5} =$

22) $1\frac{1}{18} \div 1\frac{2}{9} =$

23) $4\frac{2}{7} \div 1\frac{3}{10} =$

24) $7\frac{1}{3} \div 2\frac{2}{11} =$

25) $8\frac{2}{5} \div 1\frac{1}{6} =$

26) $9\frac{1}{3} \div 2\frac{1}{7} =$

Adding and Subtracting Decimals

✎ **Add and subtract decimals.**

1)
$$\begin{array}{r} 35.19 \\ -\ 24.28 \\ \hline \end{array}$$

4)
$$\begin{array}{r} 38.72 \\ -\ 21.68 \\ \hline \end{array}$$

7)
$$\begin{array}{r} 86.09 \\ -\ 35.14 \\ \hline \end{array}$$

2)
$$\begin{array}{r} 34.29 \\ +\ 42.58 \\ \hline \end{array}$$

5)
$$\begin{array}{r} 57.39 \\ +\ 26.54 \\ \hline \end{array}$$

8)
$$\begin{array}{r} 54.51 \\ +\ 32.66 \\ \hline \end{array}$$

3)
$$\begin{array}{r} 61.20 \\ +\ 33.75 \\ \hline \end{array}$$

6)
$$\begin{array}{r} 70.24 \\ -\ 42.35 \\ \hline \end{array}$$

9)
$$\begin{array}{r} 114.21 \\ -\ 88.69 \\ \hline \end{array}$$

✎ **Find the missing number.**

10) ___ $+ 2.8 = 5.4$

11) $4.1 +$ ___ $= 5.88$

12) $6.45 +$ ___ $= 8$

13) $7.25 -$ ___ $= 3.40$

14) ___ $- 2.35 = 4.25$

15) ___ $- 19.85 = 6.54$

16) $22.15 +$ ___ $= 28.95$

17) ___ $- 37.16 = 9.42$

18) ___ $+ 24.50 = 34.19$

19) $72.40 +$ ___ $= 125.20$

Multiplying and Dividing Decimals

✏ **Find the product.**

1) $0.5 \times 0.6 =$

2) $3.3 \times 0.4 =$

3) $1.28 \times 0.5 =$

4) $0.35 \times 0.6 =$

5) $1.85 \times 0.6 =$

6) $0.24 \times 0.5 =$

7) $5.25 \times 1.4 =$

8) $18.5 \times 4.6 =$

9) $15.4 \times 6.8 =$

10) $19.5 \times 2.6 =$

11) $32.2 \times 1.5 =$

12) $78.4 \times 4.5 =$

✏ **Find the quotient.**

13) $1.85 \div 10 =$

14) $74.6 \div 100 =$

15) $3.6 \div 3 =$

16) $9.6 \div 0.4 =$

17) $15.5 \div 0.5 =$

18) $32.8 \div 0.2 =$

19) $22.15 \div 1,000 =$

20) $53.55 \div 0.7 =$

21) $322.2 \div 0.2 =$

22) $50.67 \div 0.18 =$

23) $77.4 \div 0.8 =$

24) $27.93 \div 0.03 =$

Comparing Decimals

✍ **Write the correct comparison symbol (>, < or =).**

1) 0.70 ☐ 0.070

2) 0.049 ☐ 0.49

3) 5.090 ☐ 5.09

4) 2.57 ☐ 2.05

5) 9.03 ☐ 0.930

6) 6.06 ☐ 6.6

7) 7.02 ☐ 7.020

8) 3.04 ☐ 3.2

9) 3.61 ☐ 3.245

10) 0.986 ☐ 0.0986

11) 17.24 ☐ 17.240

12) 0.759 ☐ 0.81

13) 9.040 ☐ 9.40

14) 5.73 ☐ 5.213

15) 9.44 ☐ 9.404

16) 7.17 ☐ 7.170

17) 4.85 ☐ 4.085

18) 9.041 ☐ 9.40

19) 3.033 ☐ 3.030

20) 4.97 ☐ 4.970

Rounding Decimals

✏ **Round each decimal to the nearest whole number.**

1) 28.12   3) 16.22   5) 7.95

2) 6.9   4) 8.5   6) 52.7

✏ **Round each decimal to the nearest tenth.**

7) 31.761   9) 94.729   11) 13.219

8) 14.421   10) 77.89   12) 59.89

✏ **Round each decimal to the nearest hundredth.**

13) 8.428   15) 55.3786   17) 62.241

14) 23.812   16) 231.912   18) 19.447

✏ **Round each decimal to the nearest thousandth.**

19) 15.54324   21) 243.8652   23) 67.1983

20) 34.62586   22) 80.4529   24) 72.36788

Answers of Worksheets

Simplifying Fractions

1) $\frac{1}{2}$

2) $\frac{4}{5}$

3) $\frac{3}{4}$

4) $\frac{1}{2}$

5) $\frac{1}{4}$

6) $\frac{2}{3}$

7) $\frac{4}{5}$

8) $\frac{1}{4}$

9) $\frac{1}{2}$

10) $\frac{5}{7}$

11) $\frac{1}{4}$

12) $\frac{1}{2}$

13) $\frac{7}{8}$

14) $\frac{9}{10}$

15) $\frac{1}{3}$

16) $\frac{5}{14}$

17) $\frac{2}{7}$

18) $\frac{4}{27}$

19) $\frac{6}{31}$

20) $\frac{4}{9}$

21) $\frac{1}{8}$

22) C

23) A

24) B

Adding and Subtracting Fractions

1) $\frac{9}{9} = 1$

2) $\frac{9}{14}$

3) $\frac{5}{8}$

4) $1\frac{1}{10}$

5) $\frac{17}{20}$

6) $1\frac{1}{4}$

7) $1\frac{1}{5}$

8) $1\frac{1}{15}$

9) $1\frac{8}{21}$

10) $1\frac{1}{3}$

11) $1\frac{7}{30}$

12) $\frac{3}{4}$

13) $\frac{1}{6}$

14) $\frac{5}{8}$

15) $\frac{1}{6}$

16) $\frac{1}{20}$

17) $-\frac{1}{24}$

18) $\frac{3}{28}$

19) $\frac{13}{18}$

20) $\frac{7}{12}$

21) $\frac{19}{24}$

22) $-\frac{1}{15}$

23) $\frac{5}{28}$

24) $\frac{1}{2}$

25) $\frac{3}{28}$

26) $\frac{27}{40}$

27) $\frac{18}{35}$

28) $\frac{5}{16}$

29) $\frac{1}{2}$

30) $\frac{1}{18}$

Multiplying and Dividing Fractions

1) $\frac{3}{5}$

2) $\frac{1}{3}$

3) $\frac{1}{20}$

4) $\frac{1}{30}$

5) $\frac{1}{20}$

6) $\frac{1}{9}$

7) $\frac{2}{7}$

8) $\frac{1}{16}$

9) $\frac{3}{8}$

10) $\frac{1}{14}$

11) $\frac{1}{3}$

12) $\frac{1}{6}$

13) 2

14) $\frac{1}{2}$

15) $3\frac{3}{4}$

16) $\frac{2}{5}$

17) $\dfrac{3}{4}$ 21) $\dfrac{7}{10}$ 25) $\dfrac{5}{18}$ 29) $3\dfrac{3}{4}$

18) $4\dfrac{1}{2}$ 22) 1 26) $\dfrac{25}{36}$ 30) $\dfrac{20}{33}$

19) $\dfrac{26}{49}$ 23) $\dfrac{3}{5}$ 27) $\dfrac{9}{10}$

20) $\dfrac{2}{9}$ 24) $\dfrac{3}{4}$ 28) $\dfrac{8}{15}$

Adding and Subtracting Mixed Numbers

1) $5\dfrac{1}{2}$ 8) $7\dfrac{9}{10}$ 15) $3\dfrac{1}{4}$ 22) $2\dfrac{1}{16}$

2) 8 9) $11\dfrac{29}{35}$ 16) $3\dfrac{13}{15}$ 23) $4\dfrac{1}{24}$

3) $4\dfrac{1}{2}$ 10) $14\dfrac{19}{48}$ 17) $3\dfrac{1}{8}$ 24) $2\dfrac{71}{72}$

4) $4\dfrac{7}{12}$ 11) $1\dfrac{1}{2}$ 18) $2\dfrac{13}{15}$ 25) $5\dfrac{33}{35}$

5) $6\dfrac{5}{12}$ 12) $2\dfrac{1}{5}$ 19) $3\dfrac{7}{30}$ 26) $6\dfrac{32}{63}$

6) $8\dfrac{13}{15}$ 13) $1\dfrac{2}{9}$ 20) $4\dfrac{3}{35}$

7) $6\dfrac{16}{21}$ 14) $1\dfrac{9}{14}$ 21) $5\dfrac{1}{12}$

Multiplying and Dividing Mixed Numbers

1) $12\dfrac{3}{8}$ 10) 11 19) $4\dfrac{4}{27}$

2) $23\dfrac{1}{9}$ 11) $\dfrac{4}{7}$ 20) $2\dfrac{26}{33}$

3) $35\dfrac{15}{16}$ 12) $1\dfrac{1}{4}$ 21) $3\dfrac{17}{36}$

4) $8\dfrac{2}{3}$ 13) $4\dfrac{2}{9}$ 22) $\dfrac{19}{22}$

5) 5 14) $2\dfrac{23}{24}$ 23) $3\dfrac{27}{91}$

6) $6\dfrac{6}{7}$ 15) 3 24) $3\dfrac{13}{36}$

7) $30\dfrac{1}{3}$ 16) $\dfrac{11}{45}$ 25) $7\dfrac{1}{5}$

8) $7\dfrac{6}{7}$ 17) $\dfrac{19}{84}$ 26) $4\dfrac{16}{45}$

9) $21\dfrac{7}{8}$ 18) $\dfrac{52}{57}$

Adding and Subtracting Decimals

1) 10.91 2) 76.87 3) 94.95 4) 17.04

5) 83.93	9) 25.52	13) 3.85	17) 46.58
6) 27.89	10) 2.6	14) 6.6	18) 9.69
7) 50.95	11) 1.78	15) 26.39	19) 52.8
8) 87.17	12) 1.55	16) 6.8	

Multiplying and Dividing Decimals

1) 0.3	7) 7.35	13) 0.185	19) 0.02215
2) 1.32	8) 85.1	14) 0.746	20) 76.5
3) 0.64	9) 104.72	15) 1.2	21) 1,611
4) 0.21	10) 50.7	16) 24	22) 281.5
5) 1.11	11) 48.3	17) 31	23) 96.75
6) 0.12	12) 352.8	18) 164	24) 931

Comparing Decimals

1) >	6) <	11) =	16) =
2) <	7) =	12) <	17) >
3) =	8) <	13) <	18) <
4) >	9) >	14) >	19) >
5) >	10) >	15) >	20) =

Rounding Decimals

1) 28	9) 94.7	17) 62.24
2) 7	10) 77.9	18) 19.45
3) 16	11) 13.2	19) 15.543
4) 9	12) 59.9	20) 34.626
5) 8	13) 8.43	21) 243.865
6) 53	14) 23.81	22) 80.453
7) 31.8	15) 55.38	23) 67.198
8) 14.4	16) 231.91	24) 72.368

Chapter 3 :

Proportions, Ratios, and Percent

Topics that you will practice in this chapter:

- ✓ Simplifying Ratios
- ✓ Proportional Ratios
- ✓ Similarity and Ratios
- ✓ Ratio and Rates Word Problems
- ✓ Percentage Calculations
- ✓ Percent Problems
- ✓ Discount, Tax and Tip
- ✓ Percent of Change
- ✓ Simple Interest

Without mathematics, there's nothing you can do. Everything around you is mathematics. Everything around you is numbers." – Shakuntala Devi

Simplifying Ratios

✍ **Reduce each ratio.**

1) $15:20 =$ ___: ___

2) $7:70 =$ ___: ___

3) $16:28 =$ ___: ___

4) $7:21 =$ ___: ___

5) $4:40 =$ ___: ___

6) $6:48 =$ ___: ___

7) $16:64 =$ ___: ___

8) $10:25 =$ ___: ___

9) $8:48 =$ ___: ___

10) $49:63 =$ ___: ___

11) $18:27 =$ ___: ___

12) $35:10 =$ ___: ___

13) $90:9 =$ ___: ___

14) $24:32 =$ ___: ___

15) $7:56 =$ ___: ___

16) $45:63 =$ ___: ___

17) $56:72 =$ ___: ___

18) $26:13 =$ ___: ___

19) $15:45 =$ ___: ___

20) $28:4 =$ ___: ___

21) $24:48 =$ ___: ___

22) $30:24 =$ ___: ___

23) $70:140 =$ ___: ___

24) $6:180 =$ ___: ___

✍ **Write each ratio as a fraction in simplest form.**

25) $6:12 =$

26) $30:50 =$

27) $15:35 =$

28) $9:27 =$

29) $8:24 =$

30) $18:84 =$

31) $7:14 =$

32) $7:35 =$

33) $40:96 =$

34) $12:54 =$

35) $44:52 =$

36) $12:27 =$

37) $15:180 =$

38) $39:143 =$

39) $20:300 =$

40) $30:120 =$

41) $56:42 =$

42) $26:130 =$

43) $66:123 =$

44) $70:630 =$

45) $75:125 =$

Proportional Ratios

✍ **Fill in the blanks; Calculate each proportion.**

1) $3:8 = \underline{\quad} : 48$

2) $2:5 = 20:\underline{\quad}$

3) $1:9 = \underline{\quad} : 81$

4) $6:7 = 12:\underline{\quad}$

5) $9:2 = 63:\underline{\quad}$

6) $8:7 = \underline{\quad} : 49$

7) $20:3 = \underline{\quad} : 15$

8) $1:3 = \underline{\quad} : 75$

9) $7:6 = \underline{\quad} : 60$

10) $8:5 = \underline{\quad} : 45$

11) $3:10 = 60:\underline{\quad}$

12) $6:11 = 42:\underline{\quad}$

✍ **State if each pair of ratios form a proportion.**

13) $\frac{3}{20}$ and $\frac{9}{60}$

14) $\frac{1}{7}$ and $\frac{6}{42}$

15) $\frac{3}{7}$ and $\frac{24}{56}$

16) $\frac{4}{9}$ and $\frac{12}{18}$

17) $\frac{1}{9}$ and $\frac{12}{81}$

18) $\frac{7}{8}$ and $\frac{21}{28}$

19) $\frac{9}{13}$ and $\frac{27}{39}$

20) $\frac{1}{8}$ and $\frac{8}{64}$

21) $\frac{6}{19}$ and $\frac{30}{85}$

22) $\frac{5}{9}$ and $\frac{40}{81}$

23) $\frac{9}{14}$ and $\frac{108}{168}$

24) $\frac{15}{23}$ and $\frac{360}{552}$

✍ **Calculate each proportion.**

25) $\frac{20}{25} = \frac{32}{x}, x = \underline{\quad}$

26) $\frac{1}{8} = \frac{32}{x}, x = \underline{\quad}$

27) $\frac{15}{5} = \frac{21}{x}, x = \underline{\quad}$

28) $\frac{1}{7} = \frac{x}{294}, x = \underline{\quad}$

29) $\frac{7}{9} = \frac{x}{81}, x = \underline{\quad}$

30) $\frac{1}{5} = \frac{13}{x}, x = \underline{\quad}$

31) $\frac{9}{5} = \frac{36}{x}, x = \underline{\quad}$

32) $\frac{6}{13} = \frac{48}{x}, x = \underline{\quad}$

33) $\frac{5}{8} = \frac{x}{88}, x = \underline{\quad}$

34) $\frac{4}{15} = \frac{x}{240}, x = \underline{\quad}$

35) $\frac{9}{19} = \frac{x}{266}, x = \underline{\quad}$

36) $\frac{7}{15} = \frac{x}{270}, x = \underline{\quad}$

Similarity and Ratios

✎ **Each pair of figures is similar. Find the missing side.**

1)
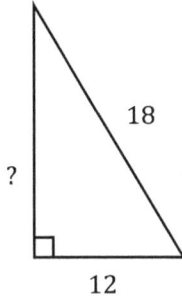

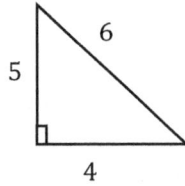

2)
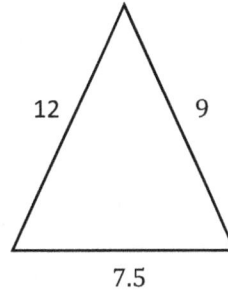

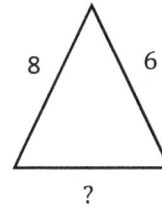

3)
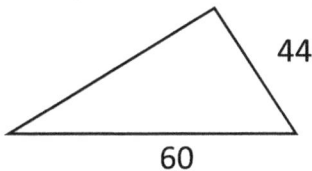

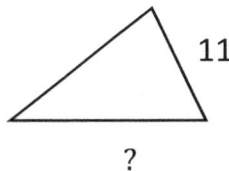

4)

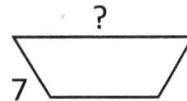

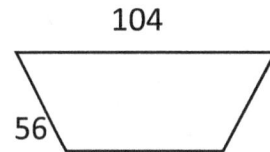

✎ **Calculate.**

5) Two rectangles are similar. The first is 24 feet wide and 120 feet long. The second is 30 feet wide. What is the length of the second rectangle? _____

6) Two rectangles are similar. One is 5 meters by 36 meters. The longer side of the second rectangle is 90 meters. What is the other side of the second rectangle? _____

7) A building casts a shadow 25 ft long. At the same time a girl 10 ft tall casts a shadow 5 ft long. How tall is the building? _____

8) The scale of a map of Texas is 4 inches: 32 miles. If you measure the distance from Dallas to Martin County as 38.4 inches, approximately how far is Martin County from Dallas? _____

Ratio and Rates Word Problems

✑ **Find the answer for each word problem.**

1) Mason has 24 red cards and 36 green cards. What is the ratio of Mason 's red cards to his green cards? _____

2) In a party, 45 soft drinks are required for every 54 guests. If there are 378 guests, how many soft drinks is required? _____

3) In Mason's class, 42 of the students are tall and 24 are short. In Michael's class 84 students are tall and 48 students are short. Which class has a higher ratio of tall to short students? _____

4) The price of 5 apples at the Quick Market is $4.6. The price of 7 of the same apples at Walmart is $5.95. Which place is the better buy? _____

5) The bakers at a Bakery can make 90 bagels in 3 hours. How many bagels can they bake in 24 hours? What is that rate per hour? _____

6) You can buy 5 cans of green beans at a supermarket for $5.75. How much does it cost to buy 45 cans of green beans? _____

7) The ratio of boys to girls in a class is 4: 7. If there are 32 boys in the class, how many girls are in that class? _____

8) The ratio of red marbles to blue marbles in a bag is 3: 7. If there are 50 marbles in the bag, how many of the marbles are red? _____

Percentage Calculations

✍ **Calculate the given percent of each value.**

1) 3% of 60 = ____

2) 20% of 32 = ____

3) 4% of 72 = ____

4) 16% of 32 = ____

5) 25% of 124 = ____

6) 35% of 56 = ____

7) 15% of 20 = ____

8) 14% of 150 = ____

9) 80% of 50 = ____

10) 12% of 115 = ____

11) 72% of 250 = ____

12) 52% of 500 = ____

13) 70% of 400 = ____

14) 27% of 145 = ____

15) 90% of 64 = ____

16) 60% of 55 = ____

17) 22% of 210 = ____

18) 8% of 235 = ____

✍ **Calculate the percent of each given value.**

19) ____% of 25 = 5

20) ____% of 40 = 20

21) ____% of 25 = 2

22) ____% of 50 = 16

23) ____% of 250 = 5

24) ____% of 40 = 32

25) ____% of 125 = 20

26) ____% of 700 = 49

27) ____% of 350 = 49

28) ___% of 500 = 210

✍ **Calculate each percent problem.**

29) A Cinema has 250 seats. 60 seats were sold for the current movie. What percent of seats are empty? _____ %

30) There are 68 boys and 92 girls in a class. 75% of the students in the class take the bus to school. How many students do not take the bus to school? ____

Percent Problems

✍ Calculate each problem.

1) 9 is what percent of 45? ____%

2) 60 is what percent of 120? ____%

3) 10 is what percent of 200? ____%

4) 15 is what percent of 125? ____%

5) 10 is what percent of 400? ____%

6) 66 is what percent of 55? ____%

7) 40 is what percent of 160? ____%

8) 40 is what percent of 50? ____%

9) 120 is what percent of 800? ____%

10) 78 is what percent of 120? ___%

11) 36 is what percent of 144? ___%

12) 17 is what percent of 85? ___%

13) 90 is what percent of 900? ___%

14) 36 is what percent of 16? ___%

15) 63 is what percent of 14? ___%

16) 18 is what percent of 60? ___%

17) 126 is what percent of 200? ___%

18) 232 is what percent of 40? ___%

✍ Calculate each percent word problem.

19) There are 40 employees in a company. On a certain day, 25 were present. What percent showed up for work? _____%

20) A metal bar weighs 60 ounces. 25% of the bar is gold. How many ounces of gold are in the bar? _____

21) A crew is made up of 12 women; the rest are men. If 15% of the crew are women, how many people are in the crew? _____

22) There are 40 students in a class and 8 of them are girls. What percent are boys? _____%

23) The Royals softball team played 400 games and won 280 of them. What percent of the games did they lose? _____%

Discount, Tax and Tip

✎ **Find the selling price of each item.**

1) Original price of a computer: $420

   Tax: 8% Selling price: $_____

2) Original price of a laptop: $280

   Tax: 4% Selling price: $_____

3) Original price of a sofa: $820

   Tax: 5% Selling price: $_____

4) Original price of a car: $15,800

   Tax: 3.6%   Selling price: $_____

5) Original price of a Table: $250

   Tax: 9% Selling price: $_____

6) Original price of a house: $630,000

   Tax: 1.8%   Selling price: $_____

7) Original price of a tablet: $450

   Discount: 30%   Selling price: $____

8) Original price of a chair: $390

   Discount: 8%   Selling price: $____

9) Original price of a book: $75

   Discount: 42%   Selling price: $____

10) Original price of a cellphone: $820

   Discount: 23%   Selling price: $___

11) Food bill: $45

   Tip: 15%   Price: $_____

12) Food bill: $32

   Tipp: 20%   Price: $_____

13) Food bill: $90

   Tip: 35%   Price: $_____

14) Food bill: $42

   Tipp: 12%   Price: $_____

✎ **Find the answer for each word problem.**

15) Nicolas hired a moving company. The company charged $500 for its services, and Nicolas gives the movers a 40% tip. How much does Nicolas tip the movers? $_____

16) Mason has lunch at a restaurant and the cost of his meal is $90. Mason wants to leave a 25% tip. What is Mason's total bill including tip? $_____

17) The sales tax in Texas is 19.80% and an item costs $350. How much is the tax? $_____

18) The price of a table at Best Buy is $680. If the sales tax is 5%, what is the final price of the table including tax? $_____

Percent of Change

✍ **Find each percent of change.**

1) From 150 to 450. ___ %

2) From 50 ft to 250 ft. ___ %

3) From $60 to $360. ___ %

4) From 60 cm to 180 cm. ___ %

5) From 15 to 45. ___ %

6) From 80 to 16. ___ %

7) From 120 to 360. ___ %

8) From 900 to 450. ___ %

9) From 1,000 to 200. ___ %

10) From 144 to 36. ___ %

✍ **Calculate each percent of change word problem.**

11) Bob got a raise, and his hourly wage increased from $42 to $63. What is the percent increase? ____ %

12) The price of a pair of shoes increases from $50 to $61. What is the percent increase? ___ %

13) At a coffee shop, the price of a cup of coffee increased from $4.80 to $5.76. What is the percent increase in the cost of the coffee? _____ %

14) 51 cm are cut from 85 cm board. What is the percent decrease in length? _____ %

15) In a class, the number of students has been increased from 54 to 81. What is the percent increase? _____ %

16) The price of gasoline rises from $24.40 to $30.50 in one month. By what percent did the gas price rise? _____ %

17) A shirt was originally priced at $38. It went on sale for $24.70. What was the percent that the shirt was discounted? _____ %

Simple Interest

✍ Determine the simple interest for these loans.

1) $480 at 11% for 3 years. $ _____

2) $4,200 at 7% for 4 years. $ _____

3) $2,500 at 20% for 3 years. $ _____

4) $6,800 at 3.9% for 4 months. $ ____

5) $800 at 6% for 7 months. $ _____

6) $36,000 at 4.2% for 6 years. $ _____

7) $6,500 at 7% for 4 years. $ _____

8) $850 at 9.5% for 2 years. $ _____

9) $1,200 at 5.8% for 9 months. $ ____

10) $3,000 at 4.5% for 7 years. $ _____

✍ Calculate each simple interest word problem.

11) A new car, valued at $22,000, depreciates at 8.5% per year. What is the value of the car one year after purchase? $_____

12) Sara puts $9,000 into an investment yielding 6% annual simple interest; she left the money in for three years. How much interest does Sara get at the end of those three years? $_____

13) A bank is offering 12% simple interest on a savings account. If you deposit $16,400, how much interest will you earn in two years? $_____

14) $720 interest is earned on a principal of $6,000 at a simple interest rate of 4% interest per year. For how many years was the principal invested? _____

15) In how many years will $2,200 yield an interest of $440 at 4% simple interest? _____

16) Jim invested $8,000 in a bond at a yearly rate of 4.5%. He earned $1,440 in interest. How long was the money invested? _____

Answers of Worksheets

Simplifying Ratios

1) 3 : 4
2) 1 : 10
3) 4 : 7
4) 1 : 3
5) 1 : 10
6) 1 : 8
7) 2 : 8
8) 2 : 5
9) 1 : 6
10) 7 : 9
11) 2 : 3
12) 7 : 2
13) 10 : 1

14) 3 : 4
15) 1 : 8
16) 5 : 7
17) 7 : 9
18) 2 : 1
19) 1 : 3
20) 7 : 1
21) 1 : 2
22) 5 : 4
23) 1 : 2
24) 1 : 30
25) $\frac{1}{2}$

26) $\frac{3}{5}$
27) $\frac{3}{7}$
28) $\frac{1}{3}$
29) $\frac{1}{3}$
30) $\frac{3}{14}$
31) $\frac{1}{2}$
32) $\frac{1}{5}$
33) $\frac{5}{12}$
34) $\frac{2}{9}$
35) $\frac{11}{13}$

36) $\frac{4}{9}$
37) $\frac{1}{12}$
38) $\frac{3}{11}$
39) $\frac{1}{15}$
40) $\frac{1}{4}$
41) $\frac{4}{3}$
42) $\frac{1}{5}$
43) $\frac{22}{41}$
44) $\frac{1}{9}$
45) $\frac{3}{5}$

Proportional Ratios

1) 18
2) 50
3) 9
4) 14
5) 14
6) 56
7) 100
8) 25
9) 70

10) 72
11) 200
12) 77
13) Yes
14) Yes
15) Yes
16) No
17) No
18) No

19) Yes
20) Yes
21) No
22) No
23) Yes
24) Yes
25) 40
26) 256
27) 7

28) 42
29) 63
30) 65
31) 20
32) 104
33) 55
34) 64
35) 126
36) 126

Similarity and ratios

1) 15
2) 5
3) 15

4) 13
5) 150 feet
6) 12.5 meters

7) 50 feet
8) 307.2 miles

Ratio and Rates Word Problems

1) 2 : 3

2) 315

3) The ratio for both classes is 7 to 4.   6) $51.75

4) Walmart is a better buy.   7) 56

5) 720, the rate is 30 per hour.   8) 15

Percentage Calculations

1) 1.8	11) 180	21) 8%
2) 6.4	12) 260	22) 32%
3) 2.88	13) 280	23) 2%
4) 5.12	14) 39.15	24) 80%
5) 31	15) 57.6	25) 16%
6) 19.6	16) 33	26) 7%
7) 3	17) 46.2	27) 14%
8) 21	18) 18.8	28) 42%
9) 40	19) 20%	29) 76%
10) 13.8	20) 50%	30) 40

Percent Problems

1) 20%	9) 15%	17) 63%
2) 50%	10) 65%	18) 580%
3) 5%	11) 25%	19) 62.5%
4) 12%	12) 20%	20) 15 ounces
5) 2.5%	13) 10%	21) 80
6) 120%	14) 225%	22) 80%
7) 25%	15) 450%	23) 30%
8) 80%	16) 30%	

Discount, Tax and Tip

1) $453.60	7) $315.00	13) $121.50
2) $291.20	8) $358.80	14) $47.04
3) $861.00	9) $43.50	15) $200.00
4) $16,368.80	10) $631.40	16) $112.50
5) $272.50	11) $51.75	17) $69.30
6) $641,340	12) $38.40	18) $714.00

Percent of Change

1) 200%	7) 200%	13) 20%
2) 400%	8) 50%	14) 60%
3) 500%	9) 80%	15) 50%
4) 200%	10) 75%	16) 25%
5) 200%	11) 50%	17) 35%
6) 80%	12) 22%	

Simple Interest

1) $158.40	7) $1,820.00	13) $3,936.00
2) $1,176.00	8) $161.50	14) 3 years
3) $1,500.00	9) $52.20	15) 5 years
4) $88.40	10) $945.00	16) 4 years
5) $28.00	11) $20,130.00	
6) $9,072.00	12) $1,620.00	

Chapter 4 :

Exponents and Radicals Expressions

Topics that you will practice in this chapter:

- ✓ Multiplication Property of Exponents
- ✓ Zero and Negative Exponents
- ✓ Division Property of Exponents
- ✓ Powers of Products and Quotients
- ✓ Negative Exponents and Negative Bases
- ✓ Scientific Notation
- ✓ Square Roots
- ✓ Simplifying Radical Expressions
- ✓ Simplifying Radical Expressions Involving Fractions
- ✓ Multiplying Radical Expressions
- ✓ Adding and Subtracting Radical Expressions

Mathematics is no more computation than typing is literature.

– John Allen Paulos

Multiplication Property of Exponents

✎ **Simplify and write the answer in exponential form.**

1) $4 \times 4^5 =$

2) $8^4 \times 8 =$

3) $7^3 \times 7^3 =$

4) $9^2 \times 9^2 =$

5) $2^2 \times 2^4 \times 2 =$

6) $5 \times 5^3 \times 5^3 =$

7) $4^3 \times 4^2 \times 4 \times 4 =$

8) $5x \times x =$

9) $x^3 \times x^3 =$

10) $x^7 \times x^2 =$

11) $x^4 \times x^3 \times x^2 =$

12) $10x \times 3x =$

13) $4x^3 \times 4x^3 =$

14) $7x^3 \times x =$

15) $3x^2 \times 4x^2 \times x^2 =$

16) $5x^4 \times x^4 =$

17) $2x^8 \times 2x =$

18) $6x \times x^5 =$

19) $4x^2 \times 6x^6 =$

20) $5yx^3 \times 4x =$

21) $7x^3 \times y^5x^7 =$

22) $y^2x^3 \times y^5x^4 =$

23) $3x^5 \times 4x^3y^4 =$

24) $4x^4 \times 9x^2y^5 =$

25) $5x^3y^4 \times 6x^8y^2 =$

26) $8x^3y^6 \times 4xy^3 =$

27) $2xy^5 \times 6x^3y^3 =$

28) $4x^5y^2 \times 4x^2y^8 =$

29) $7x \times 3y^8x^2 \times y^5 =$

30) $x^3 \times 2y^3x^4 \times 2y =$

31) $3yx^4 \times 3y^4x \times 3xy^3 =$

32) $6y^3 \times 2y^2x^4 \times 10yx^5 =$

Zero and Negative Exponents

✎ **Evaluate the following expressions.**

1) $1^{-5} =$

2) $4^{-1} =$

3) $0^{10} =$

4) $1^{15} =$

5) $5^{-2} =$

6) $3^{-3} =$

7) $9^{-1} =$

8) $10^{-2} =$

9) $12^{-2} =$

10) $2^{-5} =$

11) $3^{-4} =$

12) $2^{-4} =$

13) $6^{-3} =$

14) $10^{-3} =$

15) $30^{-1=}$

16) $15^{-2} =$

17) $4^{-3} =$

18) $2^{-7} =$

19) $5^{-3} =$

20) $4^{-4} =$

21) $3^{-5} =$

22) $10^{-4} =$

23) $2^{-10} =$

24) $8^{-3} =$

25) $20^{-2} =$

26) $14^{-2} =$

27) $9^{-3} =$

28) $100^{-2} =$

29) $5^{-4} =$

30) $4^{-6} =$

31) $\left(\frac{1}{4}\right)^{-3} =$

32) $\left(\frac{1}{6}\right)^{-2} =$

33) $\left(\frac{1}{7}\right)^{-2} =$

34) $\left(\frac{2}{3}\right)^{-3} =$

35) $\left(\frac{1}{13}\right)^{-2} =$

36) $\left(\frac{7}{12}\right)^{-2} =$

37) $\left(\frac{1}{6}\right)^{-3} =$

38) $\left(\frac{1}{300}\right)^{-2} =$

39) $\left(\frac{2}{9}\right)^{-2} =$

40) $\left(\frac{7}{5}\right)^{-1} =$

41) $\left(\frac{13}{23}\right)^{0} =$

42) $\left(\frac{1}{4}\right)^{-5} =$

Division Property of Exponents

✍ **Simplify.**

1) $\dfrac{5^6}{5^7} =$

2) $\dfrac{8^8}{8^6} =$

3) $\dfrac{4^5}{4} =$

4) $\dfrac{3}{3^5} =$

5) $\dfrac{x}{x^6} =$

6) $\dfrac{3 \times 3^2}{3^2 \times 3^5} =$

7) $\dfrac{9^4}{9^2} =$

8) $\dfrac{10 \times 10^9}{10^2 \times 10^7} =$

9) $\dfrac{7^5 \times 7^7}{7^4 \times 7^8} =$

10) $\dfrac{15x}{30x^6} =$

11) $\dfrac{3x^9}{4x^4} =$

12) $\dfrac{15x^8}{10x^9} =$

13) $\dfrac{42x^5}{6y^9} =$

14) $\dfrac{36y^8}{4x^4y^5} =$

15) $\dfrac{2x^7}{9x} =$

16) $\dfrac{49x^8y^6}{7x^9} =$

17) $\dfrac{48x^2}{24x^6y^{12}} =$

18) $\dfrac{30yx^5}{6yx^7} =$

19) $\dfrac{19x^7y}{38x^{12}y^4} =$

20) $\dfrac{9x^8}{63x^8} =$

21) $\dfrac{9x^{-9}}{4x^{-3}} =$

Powers of Products and Quotients

✎ **Simplify.**

1) $(4^3)^2 =$

2) $(2^3)^4 =$

3) $(2 \times 2^3)^2 =$

4) $(5 \times 5^5)^6 =$

5) $(19^4 \times 19^2)^3 =$

6) $(2^3 \times 2^4)^4 =$

7) $(5 \times 5^2)^2 =$

8) $(4^4)^4 =$

9) $(8x^5)^2 =$

10) $(3x^2y^4)^4 =$

11) $(7x^5y^2)^2 =$

12) $(5x^4y^4)^3 =$

13) $(2x^3y^3)^5 =$

14) $(10x^3y^4)^3 =$

15) $(13y^3y)^2 =$

16) $(5x^6x^4)^2 =$

17) $(6x^7y^6)^3 =$

18) $(12x^5x^7)^2 =$

19) $(2x^4 \times 2x)^4 =$

20) $(2x^4y^3)^5 =$

21) $(15x^7y^2)^2 =$

22) $(8x^3y^5)^3 =$

23) $(3x \times 2y^2)^4 =$

24) $\left(\frac{4x}{x^5}\right)^2 =$

25) $\left(\frac{x^4y^5}{x^3y^5}\right)^9 =$

26) $\left(\frac{36xy}{6x^5}\right)^3 =$

27) $\left(\frac{x^7}{x^8y^2}\right)^6 =$

28) $\left(\frac{xy^4}{x^3y^6}\right)^{-3} =$

29) $\left(\frac{5xy^8}{x^3}\right)^2 =$

30) $\left(\frac{xy^6}{2xy^3}\right)^{-4} =$

Negative Exponents and Negative Bases

✎ **Simplify.**

1) $-9^{-1} =$

2) $-9^{-2} =$

3) $-2^{-5} =$

4) $-x^{-7} =$

5) $11x^{-1} =$

6) $-8x^{-3} =$

7) $-12x^{-5} =$

8) $-9x^{-8}y^{-6} =$

9) $32x^{-5}y^{-1} =$

10) $10a^{-9}b^{-3} =$

11) $-17x^4y^{-6} =$

12) $-\dfrac{25}{x^{-5}} =$

13) $-\dfrac{13x}{a^{-7}} =$

14) $\left(-\dfrac{1}{3}\right)^{-4} =$

15) $\left(-\dfrac{3}{4}\right)^{-2} =$

16) $-\dfrac{14}{a^{-6}b^{-3}} =$

17) $-\dfrac{7x}{x^{-8}} =$

18) $-\dfrac{a^{-9}}{b^{-5}} =$

19) $-\dfrac{11}{x^{-5}} =$

20) $\dfrac{8b}{-16c^{-6}} =$

21) $\dfrac{12ab}{a^{-4}b^{-3}} =$

22) $-\dfrac{8n^{-4}}{32p^{-7}} =$

23) $\dfrac{16ab^{-6}}{-6c^{-5}} =$

24) $\left(\dfrac{10a}{5c}\right)^{-4} =$

25) $\left(-\dfrac{12x}{4yz}\right)^{-3} =$

26) $\dfrac{8ab^{-7}}{-5c^{-3}} =$

27) $\left(-\dfrac{x^4}{x^5}\right)^{-5} =$

28) $\left(-\dfrac{x^{-2}}{7x^3}\right)^{-2} =$

29) $\left(-\dfrac{x^{-4}}{x^2}\right)^{-6} =$

Scientific Notation

✍ **Write each number in scientific notation.**

1) $0.223 =$

2) $0.09 =$

3) $4.5 =$

4) $900 =$

5) $2,000 =$

6) $0.006 =$

7) $33 =$

8) $9,400 =$

9) $1,470 =$

10) $52,000 =$

11) $8,000,000 =$

12) $0.00009 =$

13) $2,158,000 =$

14) $0.0039 =$

15) $0.000075 =$

16) $4,300,000 =$

17) $130,000 =$

18) $4,000,000,000 =$

19) $0.00009 =$

20) $0.0039 =$

✍ **Write each number in standard notation.**

21) $4 \times 10^{-1} =$

22) $1.2 \times 10^{-3} =$

23) $2.7 \times 10^{5} =$

24) $6 \times 10^{-4} =$

25) $3.6 \times 10^{-3} =$

26) $5.5 \times 10^{5} =$

27) $3.2 \times 10^{4} =$

28) $3.88 \times 10^{6} =$

29) $7 \times 10^{-6} =$

30) $4.2 \times 10^{-7} =$

Square Roots

✎ **Find the value each square root.**

1) $\sqrt{16} =$ ___

2) $\sqrt{25} =$ ___

3) $\sqrt{1} =$ ___

4) $\sqrt{64} =$ ___

5) $\sqrt{0} =$ ___

6) $\sqrt{196} =$ ___

7) $\sqrt{4} =$ ___

8) $\sqrt{256} =$ ___

9) $\sqrt{36} =$ ___

10) $\sqrt{289} =$ ___

11) $\sqrt{169} =$ ___

12) $\sqrt{144} =$ ___

13) $\sqrt{100} =$ ___

14) $\sqrt{1,600} =$ ___

15) $\sqrt{2,500} =$ ___

16) $\sqrt{324} =$ ___

17) $\sqrt{529} =$ ___

18) $\sqrt{20} =$ ___

19) $\sqrt{625} =$ ___

20) $\sqrt{18} =$ ___

21) $\sqrt{50} =$ ___

22) $\sqrt{1,024} =$ ___

23) $\sqrt{160} =$ ___

24) $\sqrt{32} =$ ___

✎ **Evaluate.**

25) $\sqrt{4} \times \sqrt{25} =$ _____

26) $\sqrt{36} \times \sqrt{49} =$ _____

27) $\sqrt{6} \times \sqrt{6} =$ _____

28) $\sqrt{13} \times \sqrt{13} =$ _____

29) $2\sqrt{5} \times 3\sqrt{5} =$ _____

30) $\sqrt{12} \times \sqrt{3} =$ _____

31) $\sqrt{13} + \sqrt{13} =$ _____

32) $\sqrt{10} + 2\sqrt{10} =$ _____

33) $12\sqrt{7} - 10\sqrt{7} =$ _____

34) $4\sqrt{10} \times 2\sqrt{10} =$ _____

35) $5\sqrt{3} \times 8\sqrt{3} =$ _____

36) $6\sqrt{3} - \sqrt{12} =$ _____

Simplifying Radical Expressions

✎ **Simplify.**

1) $\sqrt{13x^2} =$

2) $\sqrt{75x^2} =$

3) $\sqrt[3]{27a} =$

4) $\sqrt{64x^5} =$

5) $\sqrt{216a} =$

6) $\sqrt[3]{63w^3} =$

7) $\sqrt{192x} =$

8) $\sqrt{125v} =$

9) $\sqrt[3]{128x^2} =$

10) $\sqrt{100x^9} =$

11) $\sqrt{16x^4} =$

12) $\sqrt[3]{500a^5} =$

13) $\sqrt{242} =$

14) $\sqrt{392p^3} =$

15) $\sqrt{8m^6} =$

16) $\sqrt{198x^3y^3} =$

17) $\sqrt{121x^5y^5} =$

18) $\sqrt{16a^6b^3} =$

19) $\sqrt{90x^5y^7} =$

20) $\sqrt[3]{64y^2x^6} =$

21) $10\sqrt{16x^4} =$

22) $6\sqrt{81x^2} =$

23) $\sqrt[3]{56x^2y^6} =$

24) $\sqrt[3]{1,000x^5y^7} =$

25) $8\sqrt{50a} =$

26) $\sqrt[4]{625x^8y} =$

27) $\sqrt{24x^4y^5r^3} =$

28) $5\sqrt{36x^4y^5z^8} =$

29) $3\sqrt[3]{343x^9y^7} =$

30) $5\sqrt{81a^5b^2c^9} =$

31) $\sqrt[4]{625x^8y^{16}} =$

Multiplying Radical Expressions

✎ **Simplify.**

1) $\sqrt{5} \times \sqrt{5} =$

2) $\sqrt{5} \times \sqrt{10} =$

3) $\sqrt{3} \times \sqrt{12} =$

4) $\sqrt{49} \times \sqrt{47} =$

5) $\sqrt{7} \times -2\sqrt{28} =$

6) $3\sqrt{15} \times \sqrt{5} =$

7) $4\sqrt{72} \times \sqrt{2} =$

8) $\sqrt{5} \times -\sqrt{49} =$

9) $\sqrt{55} \times \sqrt{11} =$

10) $7\sqrt{42} \times 2\sqrt{216} =$

11) $\sqrt{45}(5 + \sqrt{5}) =$

12) $\sqrt{13x^2} \times \sqrt{13x^3} =$

13) $-2\sqrt{27} \times \sqrt{3} =$

14) $2\sqrt{13x^4} \times \sqrt{13x^4} =$

15) $\sqrt{14x^3} \times \sqrt{7x^2} =$

16) $-8\sqrt{5x} \times \sqrt{7x^5} =$

17) $-2\sqrt{16x^5} \times 4\sqrt{8x^3} =$

18) $-4\sqrt{32}(8 + \sqrt{32}) =$

19) $\sqrt{32x}(10 - \sqrt{2x}) =$

20) $\sqrt{2x}(8\sqrt{x^5} + \sqrt{8}) =$

21) $\sqrt{20r}(5 + \sqrt{5}) =$

22) $-4\sqrt{7x} \times 3\sqrt{14x^5} =$

23) $-2\sqrt{12x} \times 3\sqrt{2x}$

24) $-\sqrt{7v^3}(-3\sqrt{42v}) =$

25) $(\sqrt{11} - 5)(\sqrt{11} + 5) =$

26) $(-3\sqrt{5} + 3)(\sqrt{5} - 4) =$

27) $(4 - 6\sqrt{3})(-6 + \sqrt{3}) =$

28) $(8 - 3\sqrt{5})(7 - \sqrt{5}) =$

29) $(-1 - \sqrt{3x})(4 + \sqrt{3x}) =$

30) $(-5 + 2\sqrt{7r})(-5 + \sqrt{7r}) =$

31) $(-\sqrt{7n} + 1)(-\sqrt{7} - 5) =$

32) $(-3 + \sqrt{3})(5 - 2\sqrt{3x}) =$

Simplifying Radical Expressions Involving Fractions

✎ **Simplify.**

1) $\dfrac{\sqrt{5}}{\sqrt{3}} =$

2) $\dfrac{\sqrt{18}}{\sqrt{45}} =$

3) $\dfrac{\sqrt{10}}{5\sqrt{2}} =$

4) $\dfrac{13}{\sqrt{3}} =$

5) $\dfrac{12\sqrt{5r}}{\sqrt{m^5}} =$

6) $\dfrac{11\sqrt{2}}{\sqrt{k}} =$

7) $\dfrac{6\sqrt{20x^3}}{\sqrt{16x}} =$

8) $\dfrac{\sqrt{14x^3y^4}}{\sqrt{7x^4y^3}} =$

9) $\dfrac{1}{1-\sqrt{5}} =$

10) $\dfrac{1-8\sqrt{a}}{\sqrt{11a}} =$

11) $\dfrac{\sqrt{a}}{\sqrt{a}+\sqrt{b}} =$

12) $\dfrac{1-\sqrt{5}}{2-\sqrt{6}} =$

13) $\dfrac{4+\sqrt{7}}{3-\sqrt{8}} =$

14) $\dfrac{5}{-3-3\sqrt{3}} =$

15) $\dfrac{7}{2-\sqrt{5}} =$

16) $\dfrac{\sqrt{7}-\sqrt{3}}{\sqrt{3}-\sqrt{7}} =$

17) $\dfrac{\sqrt{5}+\sqrt{7}}{\sqrt{7}-\sqrt{5}} =$

18) $\dfrac{2\sqrt{2}-\sqrt{3}}{3\sqrt{2}+\sqrt{5}} =$

19) $\dfrac{\sqrt{11}+5\sqrt{3}}{4-\sqrt{11}} =$

20) $\dfrac{\sqrt{5}+\sqrt{3}}{2-\sqrt{3}} =$

21) $\dfrac{\sqrt{32a^7b^4}}{\sqrt{2ab^3}} =$

22) $\dfrac{10\sqrt{21x^5}}{5\sqrt{x^3}} =$

Adding and Subtracting Radical Expressions

✎ **Simplify.**

1) $\sqrt{2} + \sqrt{8} =$

2) $3\sqrt{50} + 4\sqrt{2} =$

3) $2\sqrt{12} - 4\sqrt{3} =$

4) $5\sqrt{32} - 5\sqrt{2} =$

5) $3\sqrt{75} - 5\sqrt{3} =$

6) $-\sqrt{72} - 4\sqrt{2} =$

7) $-7\sqrt{16} - 4\sqrt{25} =$

8) $8\sqrt{24} + 2\sqrt{6} =$

9) $10\sqrt{49} - 7\sqrt{100} =$

10) $-7\sqrt{5} + 9\sqrt{45} =$

11) $-15\sqrt{12} + 14\sqrt{48} =$

12) $20\sqrt{4} - 2\sqrt{25} =$

13) $-2\sqrt{20} + 7\sqrt{5} =$

14) $8\sqrt{7} - 2\sqrt{63} =$

15) $5\sqrt{44} + 3\sqrt{11} =$

16) $3\sqrt{27} - 5\sqrt{48} =$

17) $\sqrt{144} - \sqrt{81} =$

18) $3\sqrt{20} - 6\sqrt{5} =$

19) $-2\sqrt{7} + 8\sqrt{28} =$

20) $3\sqrt{75} - 2\sqrt{3} =$

21) $5\sqrt{27} - 3\sqrt{3} =$

22) $-7\sqrt{30} + 6\sqrt{120} =$

23) $-7\sqrt{24} - 2\sqrt{6} =$

24) $-\sqrt{32x} + 4\sqrt{2x} =$

25) $\sqrt{7y^2} + y\sqrt{112} =$

26) $\sqrt{45mn^2} + 2n\sqrt{5m} =$

27) $-4\sqrt{12a} - 4\sqrt{3a} =$

28) $-5\sqrt{15ab} - 2\sqrt{60ab} =$

29) $\sqrt{45x^2y} + x\sqrt{20y} =$

30) $2\sqrt{7a} + 4\sqrt{63a} =$

Answers of Worksheets

Multiplication Property of Exponents

1) 4^6
2) 8^5
3) 7^6
4) 9^4
5) 2^7
6) 5^7
7) 4^7
8) $5x^2$

9) x^6
10) x^9
11) x^9
12) $30x^2$
13) $16x^6$
14) $7x^4$
15) $12x^6$
16) $5x^8$

17) $4x^9$
18) $6x^6$
19) $24x^8$
20) $20x^4y$
21) $7x^{10}y^5$
22) x^7y^7
23) $12x^8y^4$
24) $36x^6y^5$

25) $30x^{11}y^6$
26) $32x^4y^9$
27) $12x^4y^8$
28) $16x^7y^{10}$
29) $21x^3y^{13}$
30) $4x^7y^4$
31) $27x^6y^8$
32) $120x^9y^6$

Zero and Negative Exponents

1) 1
2) $\frac{1}{4}$
3) 0
4) 1
5) $\frac{1}{25}$
6) $\frac{1}{27}$
7) $\frac{1}{9}$
8) $\frac{1}{100}$
9) $\frac{1}{144}$
10) $\frac{1}{32}$
11) $\frac{1}{81}$

12) $\frac{1}{16}$
13) $\frac{1}{216}$
14) $\frac{1}{1,000}$
15) $\frac{1}{30}$
16) $\frac{1}{225}$
17) $\frac{1}{64}$
18) $\frac{1}{128}$
19) $\frac{1}{125}$
20) $\frac{1}{256}$
21) $\frac{1}{243}$

22) $\frac{1}{10,000}$
23) $\frac{1}{1,024}$
24) $\frac{1}{512}$
25) $\frac{1}{400}$
26) $\frac{1}{196}$
27) $\frac{1}{729}$
28) $\frac{1}{10,000}$
29) $\frac{1}{625}$
30) $\frac{1}{4,096}$
31) 64
32) 36

33) 49
34) $\frac{27}{8}$
35) 169
36) $\frac{144}{49}$
37) 216
38) $90,000$
39) $\frac{81}{4}$
40) $\frac{5}{7}$
41) 1
42) $1,024$

Division Property of Exponents

1) $\frac{1}{5}$
2) 8^2
3) 4^4
4) $\frac{1}{3^4}$

5) $\frac{1}{x^5}$
6) $\frac{1}{3^4}$
7) 9^2
8) 10

9) 1
10) $\frac{1}{2x^5}$
11) $\frac{3x^5}{4}$
12) $\frac{3}{2x}$

13) $\frac{7x^5}{y^9}$
14) $\frac{9y^3}{x^4}$
15) $\frac{2x^6}{9}$

16) $\frac{7y^6}{x}$

17) $\frac{2}{x^4 y^{12}}$

18) $\frac{5}{x^2}$

19) $\frac{1}{2x^5 y^3}$

20) $\frac{1}{7}$

21) $\frac{9}{4x^6}$

Powers of Products and Quotients

1) 4^6

2) 2^{12}

3) 2^8

4) 5^{36}

5) 19^{18}

6) 2^{28}

7) 5^6

8) 4^{16}

9) $64x^{10}$

10) $81x^8 y^{16}$

11) $49x^{10} y^4$

12) $125x^{12} y^{12}$

13) $32x^{15} y^{15}$

14) $1,000x^9 y^{12}$

15) $169y^8$

16) $25x^{20}$

17) $216x^{21} y^{18}$

18) $144x^{24}$

19) $256x^{20}$

20) $32x^{20} y^{15}$

21) $225x^{14} y^4$

22) $512x^9 y^{15}$

23) $1,296x^4 y^8$

24) $\frac{16}{x^8}$

25) x^9

26) $\frac{216y^3}{x^{12}}$

27) $\frac{1}{x^6 y^{12}}$

28) $x^6 y^6$

29) $\frac{25y^{16}}{x^4}$

30) $\frac{16}{y^{12}}$

Negative Exponents and Negative Bases

1) $-\frac{1}{9}$

2) $-\frac{1}{81}$

3) $-\frac{1}{32}$

4) $-\frac{1}{x^7}$

5) $\frac{11}{x}$

6) $-\frac{8}{x^3}$

7) $-\frac{12}{x^5}$

8) $-\frac{9}{x^8 y^6}$

9) $\frac{32}{x^5 y}$

10) $\frac{10}{a^9 b^3}$

11) $-\frac{17x^4}{y^6}$

12) $-25x^5$

13) $-13xa^7$

14) 81

15) $\frac{16}{9}$

16) $-14a^6 b^3$

17) $-7x^9$

18) $-\frac{b^5}{a^9}$

19) $-11x^5$

20) $-\frac{bc^6}{2}$

21) $12a^5 b^4$

22) $-\frac{p^7}{4n^4}$

23) $-\frac{8ac^5}{3b^6}$

24) $\frac{c^4}{16a^4}$

25) $\frac{y^3 z^3}{27x^3}$

26) $-\frac{8ac^3}{5b^7}$

27) $-x^5$

28) $49x^{10}$

29) x^{36}

Scientific Notation

1) 2.23×10^{-1}
2) 9×10^{-2}
3) 4.5×10^{0}
4) 9×10^{2}
5) 2×10^{3}
6) 6×10^{-3}
7) 3.3×10^{1}
8) 9.4×10^{3}
9) 1.47×10^{3}
10) 5.2×10^{4}

11) 8×10^{6}
12) 9×10^{-5}
13) 2.158×10^{6}
14) 3.9×10^{-3}
15) 7.5×10^{-5}
16) 4.3×10^{6}
17) 1.3×10^{5}
18) 4×10^{9}
19) 9×10^{-5}
20) 3.9×10^{-3}

21) 0.4
22) 0.0012
23) 270,000
24) 0.0006
25) 0.0036
26) 550,000
27) 32,000
28) 3,880,000
29) 0.000007
30) 0.00000042

Square Roots

1) 4
2) 5
3) 1
4) 8
5) 0
6) 14
7) 2
8) 16
9) 6

10) 17
11) 13
12) 12
13) 10
14) 40
15) 50
16) 18
17) 23
18) $2\sqrt{5}$

19) 25
20) $3\sqrt{2}$
21) $5\sqrt{2}$
22) 32
23) $4\sqrt{10}$
24) $4\sqrt{2}$
25) 10
26) 42
27) 6

28) 13
29) 30
30) 6
31) $2\sqrt{13}$
32) $3\sqrt{10}$
33) $2\sqrt{7}$
34) 80
35) 120
36) $4\sqrt{3}$

Simplifying radical expressions

1) $x\sqrt{13}$
2) $5x\sqrt{3}$
3) $3\sqrt[3]{a}$
4) $8x^{2}\sqrt{x}$
5) $6\sqrt{6a}$
6) $w\sqrt[3]{63}$
7) $8\sqrt{3x}$
8) $5\sqrt{5v}$

9) $4\sqrt[3]{2x^{2}}$
10) $10x^{4}\sqrt{x}$
11) $4x^{2}$
12) $5a\sqrt[3]{4a^{2}}$
13) $11\sqrt{2}$
14) $14p\sqrt{2p}$
15) $2m^{3}\sqrt{2}$
16) $3x.y\sqrt{22xy}$

17) $11x^{2}y^{2}\sqrt{xy}$
18) $4a^{3}b\sqrt{b}$
19) $3x^{2}y^{3}\sqrt{10xy}$
20) $4x^{2}\sqrt[3]{y^{2}}$
21) $40x^{2}$
22) $54x$
23) $2y^{2}\sqrt[3]{7x^{2}}$
24) $10xy^{2}\sqrt[3]{x^{2}y}$

25) $40\sqrt{2a}$

26) $5x^2\sqrt[4]{y}$

27) $2x^2y^2r\sqrt{6yr}$

28) $30x^2y^2z^4\sqrt{y}$

29) $21x^3y^2\sqrt[3]{y}$

30) $45a^2bc^4\sqrt{ac}$

31) $5x^2y^4$

Multiplying radical expressions

1) 5

2) $5\sqrt{2}$

3) 6

4) $7\sqrt{47}$

5) -28

6) $15\sqrt{3}$

7) 48

8) $-5\sqrt{7}$

9) $11\sqrt{5}$

10) $504\sqrt{7}$

11) $15\sqrt{5}+15$

12) $13x^2\sqrt{x}$

13) -18

14) $26x^4$

15) $7x^2\sqrt{2x}$

16) $-8x^3\sqrt{35}$

17) $-64x^4\sqrt{2}$

18) $-128\sqrt{2}-128$

19) $40\sqrt{2x}-8x$

20) $8x^3\sqrt{2}+4\sqrt{x}$

21) $10\sqrt{5r}+10\sqrt{r}$

22) $-84x^3\sqrt{2}$

23) $-12\sqrt{6}x$

24) $21v^2\sqrt{6}$

25) -14

26) $15\sqrt{5}-27$

27) $40\sqrt{3}-42$

28) $71-29\sqrt{5}$

29) $-3x-5\sqrt{3x}-4$

30) $14r-15\sqrt{7r}+25$

31) $7\sqrt{n}+5\sqrt{7n}-\sqrt{7}-5$

32) $-15+6\sqrt{3x}+5\sqrt{3}-6\sqrt{x}$

Simplifying radical expressions involving fractions

1) $\frac{\sqrt{15}}{3}$

2) $\frac{9\sqrt{10}}{45}=\frac{\sqrt{10}}{5}$

3) $\frac{\sqrt{20}}{10}=\frac{\sqrt{5}}{5}$

4) $\frac{13\sqrt{3}}{3}$

5) $\frac{12\sqrt{5mr}}{m^3}$

6) $\frac{11\sqrt{2k}}{k}$

7) $3x\sqrt{5}$

8) $\frac{\sqrt{2x}}{xy}$

9) $\frac{-1-\sqrt{5}}{4}$

10) $\frac{\sqrt{11a}-8a\sqrt{11}}{11a}$

11) $\frac{a-\sqrt{ab}}{a-b}$

12) $\frac{\sqrt{30}+2\sqrt{5}-\sqrt{6}-2}{2}$

13) $12 + 8\sqrt{2} + 3\sqrt{7} + 2\sqrt{14}$

14) $-\frac{5(\sqrt{3}-1)}{6}$

15) $-14 - 7\sqrt{5}$

16) -1

17) $6 + \sqrt{35}$

18) $\frac{12 - 2\sqrt{10} - 3\sqrt{6} + \sqrt{15}}{13}$

19) $\frac{4\sqrt{11}+11+20\sqrt{3}+5\sqrt{33}}{5}$

20) $2\sqrt{5} + 3 + \sqrt{15} + 2\sqrt{3}$

21) $4a^3\sqrt{b}$

22) $2x\sqrt{21}$

Adding and subtracting radical expressions

1) $3\sqrt{2}$

2) $19\sqrt{2}$

3) 0

4) $15\sqrt{2}$

5) $10\sqrt{3}$

6) $-10\sqrt{2}$

7) -48

8) $18\sqrt{6}$

9) 0

10) $20\sqrt{5}$

11) $26\sqrt{3}$

12) 30

13) $3\sqrt{5}$

14) $2\sqrt{7}$

15) $13\sqrt{11}$

16) $-11\sqrt{3}$

17) 3

18) 0

19) $14\sqrt{7}$

20) $13\sqrt{3}$

21) $12\sqrt{3}$

22) $5\sqrt{30}$

23) $-16\sqrt{6}$

24) 0

25) $5y\sqrt{7}$

26) $5n\sqrt{5m}$

27) $-12\sqrt{3a}$

28) $-9\sqrt{15ab}$

29) $5x\sqrt{5y}$

30) $14\sqrt{7a}$

Chapter 5 :

Algebraic Expressions

Topics that you will practice in this chapter:

- ✓ Simplifying Variable Expressions
- ✓ Simplifying Polynomial Expressions
- ✓ Translate Phrases into an Algebraic Statement
- ✓ The Distributive Property
- ✓ Evaluating One Variable Expressions
- ✓ Evaluating Two Variables Expressions
- ✓ Combining like Terms

Mathematics is, as it were, a sensuous logic, and relates to philosophy as do the arts, music, and plastic art to poetry. — *K. Shegel*

Simplifying Variable Expressions

✎ **Simplify each expression.**

1) $3(x + 5) =$

2) $(-4)(7x - 5) =$

3) $11x + 5 - 6x =$

4) $-4 - 2x^2 - 6x^2 =$

5) $7 + 13x^2 + 3 =$

6) $3x^2 + 7x + 15x^2 =$

7) $3x^2 - 12x^2 + 4x =$

8) $4x^2 - 8x - 2x =$

9) $6x + 7(3 - 4x) =$

10) $8x + 4(15x - 3) =$

11) $6(-3x - 9) - 17 =$

12) $-11x^2 - (-5x) =$

13) $2x + 7 + 5 - 8x =$

14) $7 + 6x - 11 - 5x =$

15) $27x + 8 - 13 - 5x =$

16) $(-11)(-5x + 2) - 41x =$

17) $19x - 4(4 - 2x) =$

18) $16x + 3(3x + 6) + 10 =$

19) $5(-2x - 4) - 13x =$

20) $16x - 3x(x + 10) =$

21) $17x + 5x(2 - 4x) =$

22) $5x(-4x - 7) + 20x =$

23) $25x - 19 + 4x^2 =$

24) $6x(x - 11) + 25 =$

25) $4x - 5 + 15x + 3x^2 =$

26) $-7x^2 - 11x - 9x =$

27) $10x - 9x^2 - 3x^2 - 7 =$

28) $13 + 3x^2 - 9x^2 - 21x =$

29) $22x + 10x^2 - 15x + 17 =$

30) $4x^2 + 25x + 21x^2 =$

31) $29 - 12x^2 - 23x - 4x^2 =$

32) $22x - 19x - 9x^2 + 30 =$

Simplifying Polynomial Expressions

✎ **Simplify each polynomial.**

1) $(2x^3 + 8x^2) - (11x + 3x^2) =$ _____

2) $(2x^5 + 7x^3) - (5x^3 + 11x^2) =$ _____

3) $(41x^4 + 5x^2) - (4x^2 + 20x^4) =$ _____

4) $13x - 8x^2 + 4(4x^2 + 3x^3) =$ _____

5) $(4x^3 - 22) + 5(3x^2 - 6x^3) =$ _____

6) $(4x^3 - 3x) - 5(2x^3 + x^4) =$ _____

7) $5(5x - 2x^3) - 2(8x^3 + 5x^2) =$ _____

8) $(3x^2 - 10x) - (5x^3 + 14x^2) =$ _____

9) $5x^3 - (3x^4 + 5x) + 2x^2 =$ _____

10) $11x^4 - (3x^2 + 5x) + 7x =$ _____

11) $(6x^2 - 3x^4) - (10x^4 + 3x^2) =$ _____

12) $2x^2 - 7x^3 + 19x^4 - 22x^3 =$ _____

13) $10x^2 - x^4 + 4x^4 - 32x^3 =$ _____

14) $-5x^2 + 17x^3 - 8x^2 - 6x =$ _____

15) $x^4 - 11x^5 - 30x^4 + 5x^2 =$ _____

16) $21x^3 + 13x - 5x^2 - 11x^3 =$ _____

Translate Phrases into an Algebraic Statement

✏️ **Write an algebraic expression for each phrase.**

1) 9 multiplied by x. _____

2) Subtract 11 from y. _____

3) 19 divided by x. _____

4) 38 decreased by y. _____

5) Add y to 40. _____

6) The square of 6. _____

7) x raised to the fifth power. _____

8) The sum of six and a number. _____

9) The difference between fifty–seven and y. _____

10) The quotient of nine and a number. _____

11) The quotient of the square of x and 25. _____

12) The difference between x and 6 is 19. _____

13) 10 times a reduced by the square of b. _____

14) Subtract the product of a and b from 41. _____

The Distributive Property

✍ **Use the distributive property to simply each expression.**

1) $4(1 + 2x) =$

2) $2(4 + 7x) =$

3) $3(4x - 4) =$

4) $(2x - 5)(-6) =$

5) $(-3)(x + 6) =$

6) $(4 + 3x)2 =$

7) $(-5)(8 - 3x) =$

8) $-(-5 - 7x) =$

9) $(-6x + 3)(-3) =$

10) $(-4)(x - 7) =$

11) $-(5 - 3x) =$

12) $3(9 + 4x) =$

13) $6(4 + 3x) =$

14) $(-5x + 3)2 =$

15) $(5 - 8x)(-3) =$

16) $(-12)(3x + 3) =$

17) $(5 - 3x)6 =$

18) $4(2 + 6x) =$

19) $8(7x - 3) =$

20) $(-2x + 3)4 =$

21) $(7 - 5x)(-9) =$

22) $(-10)(x - 8) =$

23) $(11 - 4x)3 =$

24) $(-6)(10x - 4) =$

25) $(3 - 9x)(-7) =$

26) $(-9)(x + 9) =$

27) $(-3 + 5x)(-7) =$

28) $(-5)(8 - 10x) =$

29) $12(4x - 8) =$

30) $(-10x + 13)(-3) =$

31) $(-8)(3x - 2) + 4(x + 5) =$

32) $(-8)(x + 4) - (6 + 5x) =$

Evaluating One Variable Expressions

✎ **Evaluate each expression using the value given.**

1) $8 - x$, $x = 5$

2) $x - 9$, $x = 5$

3) $5x + 4$, $x = 3$

4) $x - 13$, $x = -4$

5) $12 - x$, $x = 4$

6) $x + 2$, $x = 6$

7) $4x + 8$, $x = 3$

8) $x + (-7)$, $x = -8$

9) $4x + 5$, $x = 2$

10) $3x + 9$, $x = -2$

11) $15 + 3x - 7$, $x = 2$

12) $17 - 3x$, $x = 3$

13) $8x - 9$, $x = 4$

14) $5x + 4$, $x = -3$

15) $10x + 5$, $x = 3$

16) $14 - 4x$, $x = -6$

17) $3(5x + 3)$, $x = 9$

18) $4(-3x - 6)$, $x = 3$

19) $7x - 2x + 12$, $x = 4$

20) $(5x + 6) \div 2$, $x = 8$

21) $(x + 18) \div 10$, $x = 12$

22) $5x - 12 + 3x$, $x = -3$

23) $(6 - 4x)(-3)$, $x = -4$

24) $9x^2 + 3x - 6$, $x = 2$

25) $x^2 - 10x$, $x = -5$

26) $3x(7 - 2x)$, $x = 2$

27) $12x + 6 - 2x^2$, $x = -4$

28) $(-3)(4x - 8 + 3x)$, $x = 3$

29) $(-6) + \frac{x}{4} + 3x$, $x = 16$

30) $(-6) + \frac{x}{5}$, $x = 35$

31) $\left(-\frac{45}{x}\right) - 7 + 2x$, $x = 9$

32) $\left(-\frac{21}{x}\right) - 12 + 4x$, $x = 7$

Evaluating Two Variables Expressions

✎ **Evaluate each expression using the values given.**

1) $2x - 4y$,

   $x = 4, y = 1$

2) $3x + 5y$,

   $x = -2, y = 2$

3) $-7a + 4b$,

   $a = 2, b = 4$

4) $3x + 5 - y$,

   $x = 5, y = 6$

5) $3z + 12 - 2k$,

   $z = 5, k = 6$

6) $6(-x - 3y)$,

   $x = 5, y = -2$

7) $5a + 3b$,

   $a = 3, b = 4$

8) $7x \div 3y$,

   $x = 3, y = 7$

9) $2x + 15 + 5y$,

   $x = -3, y = 1$

10) $5a - (18 - b)$,

   $a = 2, b = 8$

11) $2z + 20 + 5k$,

   $z = -6, k = 5$

12) $xy + 10 + 4x$,

   $x = 3, y = 5$

13) $2x + 4y - 8 + 5$,

   $x = 5, y = 2$

14) $\left(-\frac{24}{x}\right) + 3 + 2y$,

   $x = 4, y = 6$

15) $(-3)(-3a - 3b)$,

   $a = 4, b = 5$

16) $12 + 4x - 7 - y$,

   $x = 3, y = 5$

17) $11x + 5 - 8y + 6$,

   $x = 5, y = 2$

18) $10 + 2(-4x - 5y)$,

   $x = 5, y = 4$

19) $5x + 13 + 6y$,

   $x = 5, y = 6$

20) $10a - (7a + 3b) - 11$,

   $a = 3, b = 8$

Combining like Terms

✎ **Simplify each expression.**

1) $11x + 3x + 6 =$

2) $8(2x - 6) =$

3) $18x - 7x + 11 =$

4) $(-4)(6x - 7) =$

5) $22x - 10x - 5 =$

6) $32x - 13 + 8x =$

7) $15 - (8x - 11) =$

8) $-24x + 17 - 11x =$

9) $12x - 8 - 6x + 9 =$

10) $21x + 5 - 36 + 12x =$

11) $28x + 3x - 11 =$

12) $(-3x + 4)5 =$

13) $2 + 4x + 9x - 8 =$

14) $6(2x - 5x) - 4 =$

15) $4(5x + 11) + 3x =$

16) $x - 14 - 11x =$

17) $5(10 + 9x) - 8x =$

18) $42x + 17 - 23x =$

19) $(-7x) + 19 + 20x =$

20) $(-7x) - 33 + 29x =$

21) $4(5x + 3) - 19x =$

22) $5(6 - 2x) - 15x =$

23) $-24x + (11 - 18x) =$

24) $(-9) - (6)(7x + 3) =$

25) $(-1)(8x - 10) - 21x =$

26) $-36x + 14 + 27x - 5x =$

27) $3(-13x + 6) - 17x =$

28) $-5x - 42 + 32x =$

29) $37x - 19x + 15 - 9x =$

30) $3(5x + 7x) - 31 =$

31) $14 - 6x - 15 - 9x =$

32) $-2(-5x - 7x) + 27x =$

Answers of Worksheets

Simplifying Variable Expressions

1) $3x + 15$

2) $-28x + 20$

3) $5x + 5$

4) $-8x^2 - 4$

5) $13x^2 + 10$

6) $18x^2 + 7x$

7) $-9x^2 + 4x$

8) $4x^2 - 10x$

9) $-22x + 21$

10) $68x - 12$

11) $-18x - 71$

12) $-11x^2 + 5x$

13) $-6x + 12$

14) $x - 4$

15) $22x - 5$

16) $14x - 22$

17) $27x - 16$

18) $25x + 28$

19) $-23x - 20$

20) $-3x^2 - 14x$

21) $-20x^2 + 27x$

22) $-20x^2 - 15x$

23) $4x^2 + 25x - 19$

24) $6x^2 - 66x + 25$

25) $3x^2 + 19x - 5$

26) $-7x^2 - 20x$

27) $-12x^2 + 10x - 7$

28) $-6x^2 - 21x + 13$

29) $10x^2 + 7x + 17$

30) $25x^2 + 25x$

31) $-16x^2 - 23x + 29$

32) $-9x^2 + 3x + 30$

Simplifying Polynomial Expressions

1) $2x^3 + 5x^2 - 11x$

2) $2x^5 + 2x^3 - 11x^2$

3) $21x^4 + x^2$

4) $12x^3 + 8x^2 + 13x$

5) $-26x^3 + 15x^2 - 22$

6) $-5x^4 - 6x^3 - 3x$

7) $-26x^3 - 10x^2 + 25x$

8) $-5x^3 - 11x^2 - 10x$

9) $-3x^4 + 5x^3 + 2x^2 - 5x$

10) $11x^4 - 3x^2 + 2x$

11) $-13x^4 + 3x^2$

12) $19x^4 - 29x^3 + 2x^2$

13) $3x^4 - 32x^3 + 10x^2$

14) $17x^3 - 13x^2 - 6x$

15) $-11x^5 - 29x^4 + 5x^2$

16) $10x^3 - 5x^2 + 13x$

Translate Phrases into an Algebraic Statement

1) $9x$

2) $y - 11$

3) $\frac{19}{x}$

4) $38 - y$

5) $y + 40$

6) 6^2

7) x^5

8) $6 + x$

9) $57 - y$

10) $\frac{9}{x}$

11) $\frac{x^2}{25}$

12) $x - 6 = 19$

13) $10a - b^2$

14) $41 - ab$

The Distributive Property

1) $8x + 4$

2) $14x + 8$

3) $12x - 12$

4) $-12x + 30$

5) $-3x - 18$

6) $6x + 8$

7) $15x - 40$

8) $7x + 5$

9) $18x - 9$

10) $-4x + 28$

11) $3x - 5$

12) $12x + 27$

13) $18x + 24$	18) $24x + 8$	23) $-12x + 33$	28) $50x - 40$
14) $-10x + 6$	19) $56x - 24$	24) $-60x + 24$	29) $48x - 96$
15) $24x - 15$	20) $-8x + 12$	25) $63x - 21$	30) $30x - 39$
16) $-36x - 36$	21) $45x - 63$	26) $-9x - 81$	31) $-20x + 36$
17) $-18x + 30$	22) $-10x + 80$	27) $-35x + 21$	32) $-13x - 38$

Evaluating One Variables

1) 3	9) 13	17) 144	25) 75
2) -4	10) 3	18) -60	26) 18
3) 19	11) 14	19) 32	27) -74
4) -17	12) 8	20) 23	28) -39
5) 8	13) 23	21) 3	29) 46
6) 8	14) -11	22) -36	30) 1
7) 20	15) 35	23) -66	31) 6
8) -15	16) 38	24) 36	32) 13

Evaluating Two Variables

1) 4	6) 6	11) 33	16) 12
2) 4	7) 27	12) 37	17) 50
3) 2	8) 1	13) 15	18) -70
4) 14	9) 14	14) 9	19) 74
5) 15	10) 0	15) 81	20) -26

Combining like Terms

1) $14x + 6$	9) $6x + 1$	17) $37x + 50$	25) $-29x + 10$
2) $16x - 48$	10) $33x - 31$	18) $19x + 17$	26) $-14x + 14$
3) $11x + 11$	11) $31x - 11$	19) $13x + 19$	27) $-56x + 18$
4) $-24x + 28$	12) $-15x + 20$	20) $22x - 33$	28) $27x - 42$
5) $12x - 5$	13) $13x - 6$	21) $x + 12$	29) $9x + 15$
6) $40x - 13$	14) $-18x - 4$	22) $-25x + 30$	30) $36x - 31$
7) $-8x + 26$	15) $23x + 44$	23) $-42x + 11$	31) $-15x - 1$
8) $-35x + 17$	16) $-10x - 14$	24) $-42x - 27$	32) $51x$

Chapter 6

Equations and Inequalities

Topics that you will practice in this chapter:

- ✓ One–Step Equations
- ✓ Multi–Step Equations
- ✓ Graphing Single–Variable Inequalities
- ✓ One–Step Inequalities
- ✓ Multi-Step Inequalities
- ✓ Systems of Equations
- ✓ Systems of Equations Word Problems

"Life is a math equation. In order to gain the most, you have to know how to convert negatives into positives." – Anonymous

One–Step Equations

✒ **Find the answer for each equation.**

1) $3x = 90, x =$ _____

2) $5x = 35, x =$ _____

3) $6x = 24, x =$ _____

4) $24x = 144, x =$ _____

5) $x + 15 = 20, x =$ _____

6) $x - 7 = 4, x =$ _____

7) $x - 9 = 2, x =$ _____

8) $x + 15 = 23, x =$ _____

9) $x - 4 = 13, x =$ _____

10) $12 = 16 + x, x =$ _____

11) $x - 10 = 2, x =$ _____

12) $5 - x = -11, x =$ _____

13) $28 = -6 + x, x =$ _____

14) $x - 20 = -35, x =$ _____

15) $x + 14 = -4, x =$ _____

16) $14 = 28 - x, x =$ _____

17) $7 + x = -7, x =$ _____

18) $x - 16 = 4, x =$ _____

19) $30 = x - 15, x =$ _____

20) $x - 5 = -18, x =$ _____

21) $x - 10 = 24, x =$ _____

22) $x - 20 = -25, x =$ _____

23) $x - 17 = 30, x =$ _____

24) $-70 = x - 28, x =$ _____

25) $x - 9 = 13, x =$ _____

26) $36 = 4x, x =$ _____

27) $x - 35 = 25, x =$ _____

28) $x - 25 = 10, x =$ _____

29) $70 - x = 16, x =$ _____

30) $x - 10 = 14, x =$ _____

31) $17 - x = -13, x =$ __

32) $x - 9 = -30, x =$ _____

Multi–Step Equations

✎ **Find the answer for each equation.**

1) $3x + 3 = 9$

2) $-x + 5 = 12$

3) $4x - 8 = 8$

4) $-(3 - x) = 5$

5) $4x - 8 = 16$

6) $12x - 15 = 9$

7) $2x - 18 = 2$

8) $4x + 8 = 16$

9) $24x + 27 = 75$

10) $-14(3 + x) = 14$

11) $-3(2 + x) = 6$

12) $12 = -(x - 7)$

13) $3(3 - x) = 30$

14) $-15 = -(3x + 6)$

15) $40(3 + x) = 40$

16) $5(x - 10) = 25$

17) $-18 = x + 8x$

18) $3x + 25 = -2x - 10$

19) $7(6 + 3x) = -63$

20) $18 - 3x = -4 - 5x$

21) $4 - 6x = 36 + 2x$

22) $15 + 15x = -5 + 5x$

23) $42 = (-6x) - 7 + 7$

24) $21 = 3x - 21 + 4x$

25) $-18 = -6x - 9 + 3x$

26) $5x - 15 = -29 + 6x$

27) $7x - 18 = 4x + 3$

28) $-7 - 4x = 5(4 - x)$

29) $x - 5 = -5(-3 - x)$

30) $13x - 68 = 15x - 102$

31) $-5x - 3 = -3(9 + 3x)$

32) $-2x - 15 = 6x + 17$

Graphing Single–Variable Inequalities

✎ **Draw a graph for each inequality.**

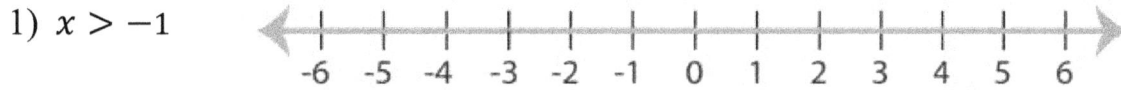

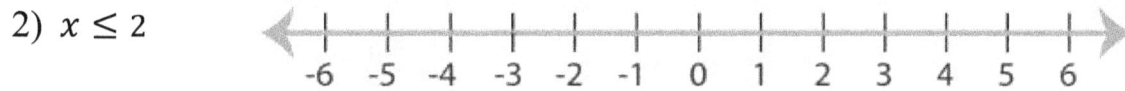

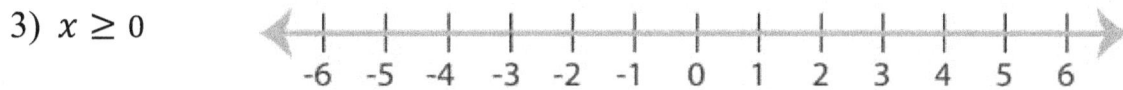

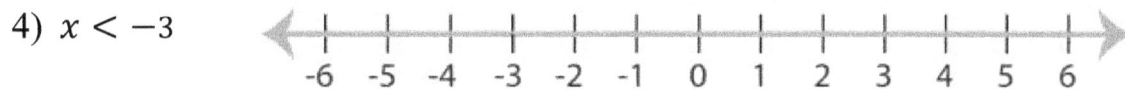

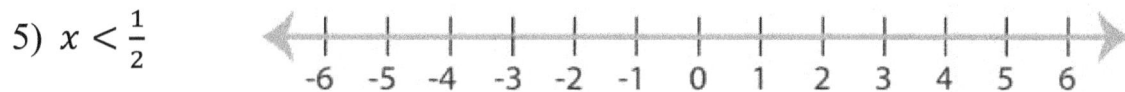

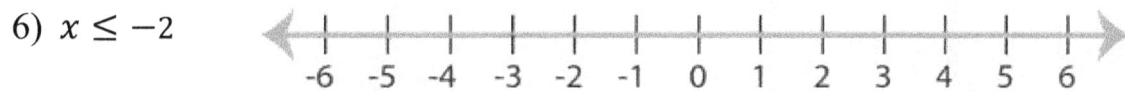

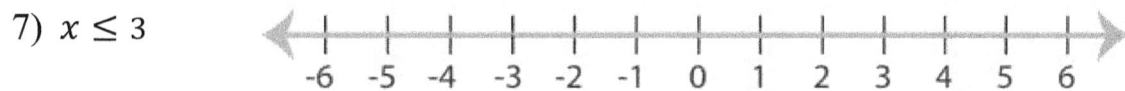

1) $x > -1$

2) $x \leq 2$

3) $x \geq 0$

4) $x < -3$

5) $x < \frac{1}{2}$

6) $x \leq -2$

7) $x \leq 3$

8) $x \geq -\frac{7}{2}$

One–Step Inequalities

✍ **Find the answer for each inequality and graph it.**

1) $x + 4 \geq 4$

2) $x - 5 \leq 2$

3) $5x > 35$

4) $9 + x \leq 11$

5) $x - 5 < -9$

6) $9x \geq 72$

7) $9x \leq 27$

8) $x + 19 > 16$

Multi-Step Inequalities

✎ **Calculate each inequality.**

1) $x - 3 \leq 7$

2) $8 - x \leq 8$

3) $3x - 9 \leq 9$

4) $4x - 4 \geq 8$

5) $x - 7 \geq 1$

6) $5x - 15 \leq 5$

7) $6x - 8 \leq 4$

8) $-11 + 6x \leq 12$

9) $4(x - 4) \leq 16$

10) $3x - 10 \leq 11$

11) $5x - 25 < 25$

12) $9x - 5 < 22$

13) $20 - 7x \geq -15$

14) $33 + 6x < 45$

15) $8 + 8x \geq 96$

16) $7 + 3x < 13$

17) $4x - 3 < 9$

18) $5(2 - 2x) \geq -30$

19) $-(7 + 6x) < 29$

20) $12 - 8x \geq -20$

21) $-4(x - 6) > 24$

22) $\dfrac{3x + 9}{6} \leq 10$

23) $\dfrac{4x - 10}{3} \leq 2$

24) $\dfrac{2x - 8}{3} > 2$

25) $8 + \dfrac{x}{6} < 9$

26) $\dfrac{9x}{7} - 4 < 5$

27) $\dfrac{15x + 45}{15} > 1$

28) $16 + \dfrac{x}{4} < 6$

Systems of Equations

✏ **Calculate each system of equations.**

1) $-x + y = 2$ $x = \underline{\quad}$
   $-4x + 2y = 6$ $y = \underline{\quad}$

2) $-15x + 3y = -9$ $x = \underline{\quad}$
   $9x - 16y = 48$ $y = \underline{\quad}$

3) $y = -7$ $x = \underline{\quad}$
   $6x + 5y = 7$ $y = \underline{\quad}$

4) $3y = -9x + 15$ $x = \underline{\quad}$
   $5x - 4y = -3$ $y = \underline{\quad}$

5) $10x - 9y = -13$ $x = \underline{\quad}$
   $-5x + 3y = 11$ $y = \underline{\quad}$

6) $-12x - 16y = 20$ $x = \underline{\quad}$
   $6x - 12y = 30$ $y = \underline{\quad}$

7) $5x - 14y = -23$ $x = \underline{\quad}$
   $-18x + 21y = 24$ $y = \underline{\quad}$

8) $15x - 21y = -6$ $x = \underline{\quad}$
   $2x - 3y = -2$ $y = \underline{\quad}$

9) $-x + 3y = 3$ $x = \underline{\quad}$
   $-14x + 16y = -10$ $y = \underline{\quad}$

10) $x + 5y = 50$ $x = \underline{\quad}$
   $3x + 10y = 80$ $y = \underline{\quad}$

11) $6x - 7y = -8$ $x = \underline{\quad}$
   $-x - 4y = -9$ $y = \underline{\quad}$

12) $2x + 4y = -10$ $x = \underline{\quad}$
   $2x - 8y = 14$ $y = \underline{\quad}$

13) $4x + 3y = 12$ $x = \underline{\quad}$
   $5x - 3y = 15$ $y = \underline{\quad}$

14) $3x - 2y = 3$ $x = \underline{\quad}$
   $7x - 8y = 22$ $y = \underline{\quad}$

15) $3x + 2y = 5$ $x = \underline{\quad}$
   $-10x - 4y = -14$ $y = \underline{\quad}$

16) $10x + 7y = 1$ $x = \underline{\quad}$
   $-5x - 7y = 24$ $y = \underline{\quad}$

Systems of Equations Word Problems

✎ **Find the answer for each word problem.**

1) Tickets to a movie cost $4 for adults and $3 for students. A group of friends purchased 8 tickets for $31.00. How many adults ticket did they buy? ____

2) At a store, Eva bought two shirts and five hats for $77.00. Nicole bought three same shirts and four same hats for $84.00. What is the price of each shirt? _____

3) A farmhouse shelters 18 animals, some are pigs, and some are ducks. Altogether there are 66 legs. How many pigs are there? _____

4) A class of 214 students went on a field trip. They took 36 vehicles, some cars and some buses. If each car holds 5 students and each bus hold 22 students, how many buses did they take? _____

5) A theater is selling tickets for a performance. Mr. Smith purchased 5 senior tickets and 3 child tickets for $105 for his friends and family. Mr. Jackson purchased 3 senior tickets and 5 child tickets for $79. What is the price of a senior ticket? $_____

6) The difference of two numbers is 10. Their sum is 20. What is the bigger number? $_____

7) The sum of the digits of a certain two–digit number is 7. Reversing its digits increase the number by 9. What is the number? _____

8) The difference of two numbers is 11. Their sum is 25. What are the numbers? _____

9) The length of a rectangle is 5 meters greater than 2 times the width. The perimeter of rectangle is 28 meters. What is the length of the rectangle? _____

10) Jim has 25 nickels and dimes totaling $1.80. How many nickels does he have? _____

Answers of Worksheets

One–Step Equations

1) 30	9) 17	17) −14	25) 22
2) 7	10) −4	18) 20	26) 9
3) 4	11) 12	19) 45	27) 60
4) 6	12) 16	20) −13	28) 35
5) 5	13) 34	21) 34	29) 54
6) 11	14) −15	22) −5	30) 24
7) 11	15) −18	23) 47	31) 30
8) 8	16) 14	24) −42	32) −21

Multi–Step Equations

1) 2	9) 2	17) −2	25) 3
2) −7	10) −4	18) −7	26) 14
3) 4	11) −4	19) −5	27) 7
4) 8	12) −5	20) −11	28) 27
5) 6	13) −7	21) −4	29) −5
6) 2	14) 3	22) −2	30) 17
7) 10	15) −2	23) −7	31) −6
8) 2	16) 15	24) 6	32) −4

Graphing Single–Variable Inequalities

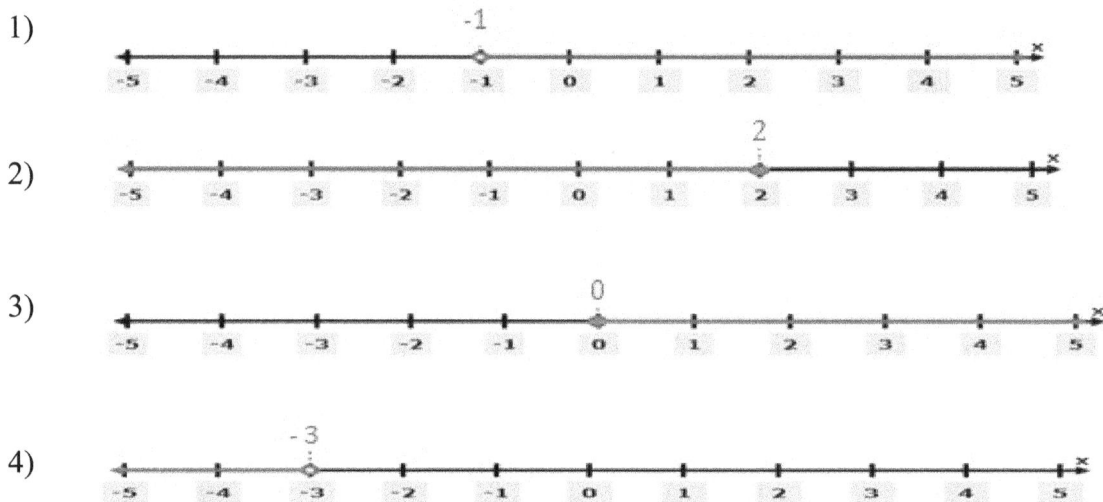

1)

2)

3)

4)

5)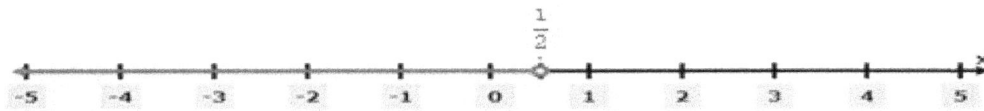

6)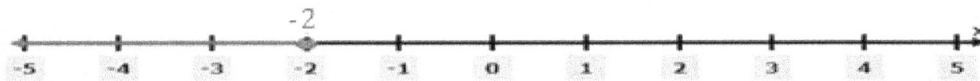

7)

8)

One–Step Inequalities

1)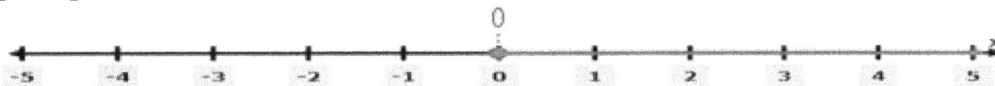

2)

3)

4)

5)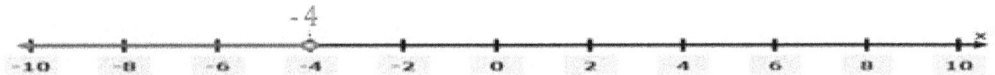

6)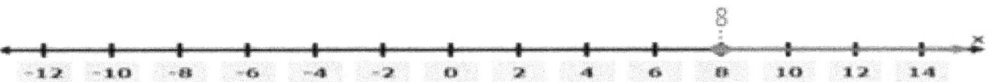

7)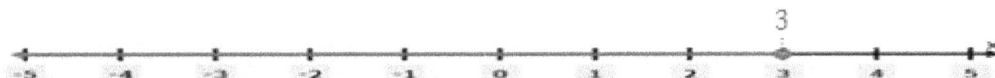

8)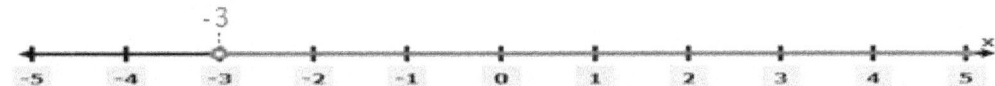

Multi-Step Inequalities

1) $x \leq 10$

2) $x \geq 0$

3) $x \leq 6$

4) $x \geq 3$

5) $x \geq 8$

6) $x \leq 4$

7) $x \leq 2$

8) $x \leq \frac{23}{6}$

9) $x \leq 8$

10) $x \leq 7$

11) $x < 10$

12) $x < 3$

13) $x \leq 5$

14) $x < 2$

15) $x \geq 11$

16) $x < 2$

17) $x < 3$

18) $x \leq 4$

19) $x > -6$

20) $x \leq 4$

21) $x < 0$

22) $x \leq 17$

23) $x \leq 4$

24) $x > 7$

25) $x < 6$ 26) $x < 7$ 27) $x > -2$ 28) $x < -40$

Systems of Equations

1) $x = -1, y = 1$ 7) $x = 1, y = 2$ 13) $x = 3, y = 0$

2) $x = 0, y = -3$ 8) $x = 8, y = 6$ 14) $x = -2, y = -\frac{9}{2}$

3) $x = 7$ 9) $x = 3, y = 2$ 15) $x = 1, y = 1$

4) $x = 1, y = 2$ 10) $x = -20, y = 14$ 16) $x = 5, y = -7$

5) $x = -4, y = -3$ 11) $x = 1, y = 2$

6) $x = 1, y = -2$ 12) $x = -1, y = -2$

Systems of Equations Word Problems

1) 7   5) $18   9) 11 meters

2) $16   6) 15   10) 14

3) 15   7) 34

4) 2   8) 18, 7

Chapter 7 :

Linear Functions

Topics that you will practice in this chapter:

- ✓ Finding Slope
- ✓ Graphing Lines Using Line Equation
- ✓ Writing Linear Equations
- ✓ Graphing Linear Inequalities
- ✓ Finding Midpoint
- ✓ Finding Distance of Two Points

"Nature is written in mathematical language." – Galileo Galilei

Finding Slope

✎ **Find the slope of each line.**

1) $y = x + 8$

2) $y = -3x + 5$

3) $y = 2x + 12$

4) $y = -4x + 19$

5) $y = 11 + 6x$

6) $y = 7 - 5x$

7) $y = 8x + 19$

8) $y = -9x + 20$

9) $y = -7x + 4$

10) $y = 3x - 8$

11) $y = \frac{1}{3}x + 8$

12) $y = -\frac{4}{5}x + 9$

13) $-3x + 6y = 30$

14) $4x + 4y = 16$

15) $3y - x = 10$

16) $8y - x = 5$

✎ **Find the slope of the line through each pair of points.**

17) $(2, 3), (7, 10)$

18) $(-3, 5), (2, 15)$

19) $(5, -3), (1, 9)$

20) $(-5, -5), (10, 25)$

21) $(22, 3), (7, 18)$

22) $(-16, 8), (-7, 26)$

23) $(25, 11), (29, 19)$

24) $(26, -19), (14, 17)$

25) $(22, -13), (20, -11)$

26) $(19, 7), (15, -3)$

27) $(5, 7), (11, 19)$

28) $(52, -62), (40, 70)$

Graphing Lines Using Line Equation

✍ **Sketch the graph of each line.**

1) $y = x - 2$

2) $y = -3x + 2$

3) $x + y = 0$

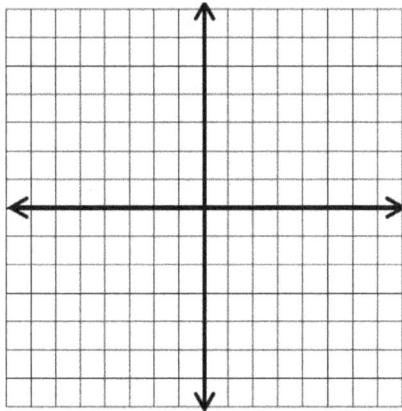
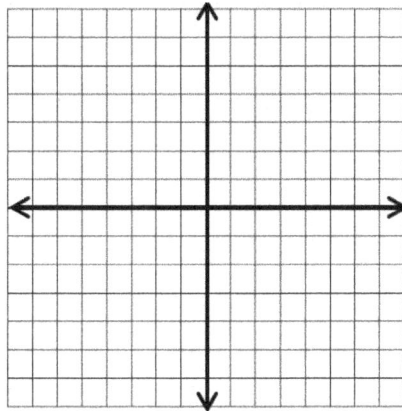
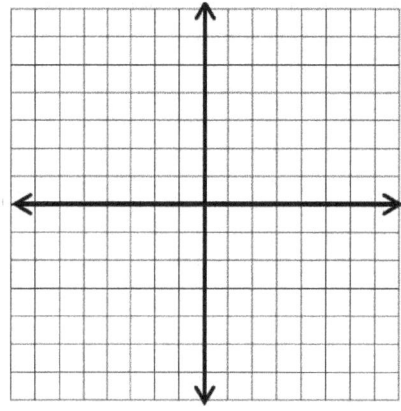

4) $x + y = -3$

5) $2x + 3y = -4$

6) $y - 3x + 6 = 0$

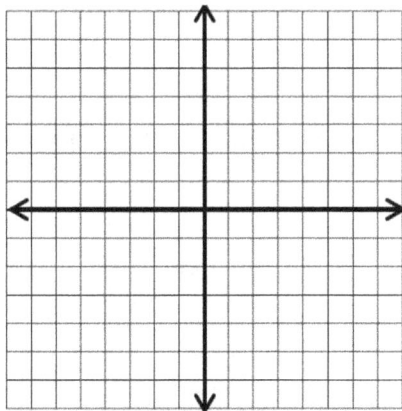
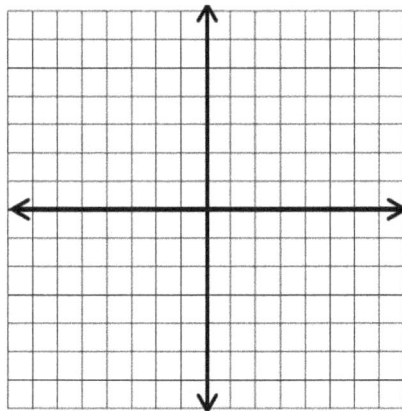
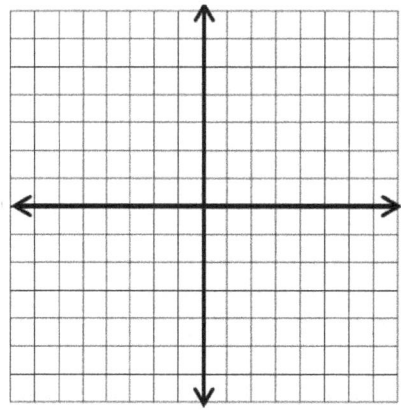

Writing Linear Equations

✍ **Write the equation of the line through the given points.**

1) Through: $(2, -5), (3, 9)$

2) Through: $(-6, 3), (3, 12)$

3) Through: $(10, 7), (5, 27)$

4) Through: $(15, 11), (3, -1)$

5) Through: $(24, 17), (12, -7)$

6) Through: $(8, 29), (4, -7)$

7) Through: $(20, -16), (12, 0)$

8) Through: $(-3, 10), (2, -5)$

9) Through: $(-6, 17), (4, -3)$

10) Through: $(-8, 22), (5, -4)$

11) Through: $(9, 27), (3, -3)$

12) Through: $(11, 32), (9, 4)$

13) Through: $(-3, 13), (-4, 0)$

14) Through: $(-5, 5), (5, 15)$

15) Through: $(18, -32), (11, 3)$

16) Through: $(-4, 25), (4, -15)$

✍ **Find the answer for each problem.**

17) What is the equation of a line with slope 6 and intercept 12?

_____

18) What is the equation of a line with slope -11 and intercept -4?

_____

19) What is the equation of a line with slope -3 and passes through point $(5, 2)$? _____

20) What is the equation of a line with slope -5 and passes through point $(-2, -1)$? _____

21) The slope of a line is -10 and it passes through point $(-3, 0)$. What is the equation of the line? _____

22) The slope of a line is 8 and it passes through point $(0, 7)$. What is the equation of the line? _____

Graphing Linear Inequalities

✎ **Sketch the graph of each linear inequality.**

1) $y > 4x - 5$ 2) $y < 2x + 4$ 3) $y \leq -5x - 2$

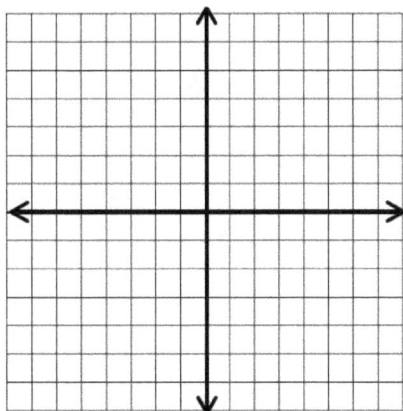   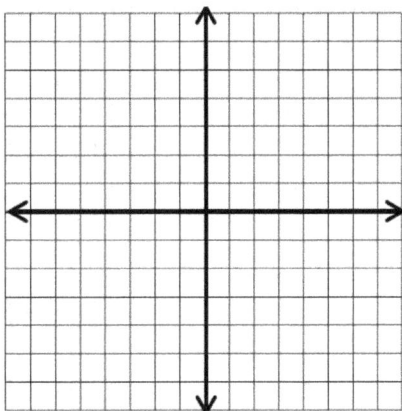   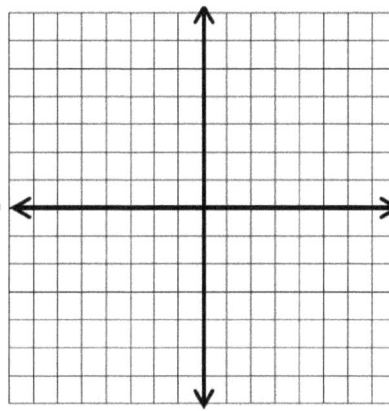

4) $4y \geq 12 + 4x$ 5) $-12y < 3x - 24$ 6) $5y \geq -15x + 10$

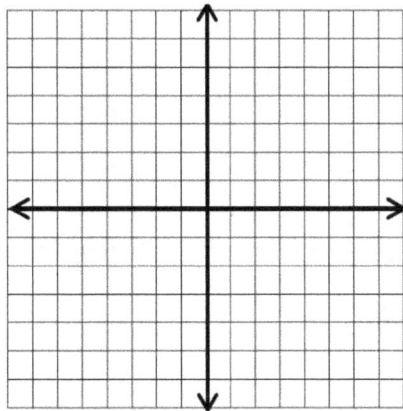   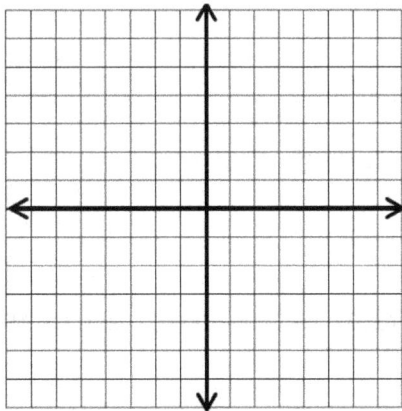   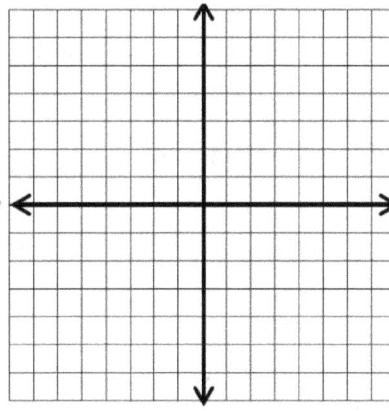

Finding Midpoint

✎ **Find the midpoint of the line segment with the given endpoints.**

1) $(-4, -3), (2, 3)$

2) $(9, 0), (-1, 8)$

3) $(9, -6), (3, 14)$

4) $(-10, -6), (0, 8)$

5) $(2, -5), (14, -15)$

6) $(-10, -3), (4, -13)$

7) $(8, 7), (-8, 13)$

8) $(-3, 6), (-9, 2)$

9) $(-4, 5), (16, -9)$

10) $(7, 14), (9, -2)$

11) $(-8, 6), (6, 6)$

12) $(10, 5), (-2, -3)$

13) $(-5, 12), (-3, 3)$

14) $(12, 7), (8, -2)$

15) $(10, 2), (-6, 14)$

16) $(-1, -2), (-7, 10)$

17) $(7, -7), (13, -13)$

18) $(-3, -8), (11, -4)$

19) $(5, -11), (-8, 9)$

20) $(14, -4), (16, 14)$

21) $(0, -5), (8, -1)$

22) $(3, 0), (-21, 18)$

23) $(17, -3), (-7, -5)$

24) $(26, -12), (6, 24)$

✎ **Find the answer for each problem.**

25) One endpoint of a line segment is $(-3, 7)$ and the midpoint of the line segment is $(-6, 9)$. What is the other endpoint? _____

26) One endpoint of a line segment is $(-3, 7)$ and the midpoint of the line segment is $(1, 5)$. What is the other endpoint? _____

27) One endpoint of a line segment is $(-10, -16)$ and the midpoint of the line segment is $(2, 9)$. What is the other endpoint? _____

Finding Distance of Two Points

✎ **Find the distance between each pair of points.**

1) $(6, 3), (-3, -9)$

2) $(5, 2), (-10, -6)$

3) $(8, 5), (8, 3)$

4) $(-8, -2), (2, 22)$

5) $(6, -7), (-3, -7)$

6) $(12, 0), (-9, -20)$

7) $(3, 20), (3, -5)$

8) $(10, 17), (5, 5)$

9) $(7, -2), (-4, -2)$

10) $(13, 4), (5, -2)$

11) $(11, 13), (5, 5)$

12) $(1, 4), (-23, -3)$

13) $(9, 8), (5, -4)$

14) $(-11, -4), (5, 8)$

15) $(-2, -6), (-2, -12)$

16) $(-1, -4), (23, 3)$

17) $(19, 3), (7, -6)$

18) $(-5, -2), (3, 4)$

19) $(2, 6), (2, -12)$

20) $(-4, -2), (8, -2)$

✎ **Find the answer for each problem.**

21) Triangle ABC is a right triangle on the coordinate system and its vertices are $(-2, 5)$, $(-2, 1)$, and $(1, 1)$. What is the area of triangle ABC? _____

22) Three vertices of a triangle on a coordinate system are $(3, -6)$, $(-5, -12)$, and $(3, -18)$. What is the perimeter of the triangle? _____

23) Four vertices of a rectangle on a coordinate system are $(-2, 2)$, $(-2, 6)$, $(4, 2)$, and $(4, 6)$. What is its perimeter? _____

Answers of Worksheets

Finding Slope

1) 1

2) −3

3) 2

4) −4

5) 6

6) −5

7) 8

8) −9

9) −7

10) 3

11) $\frac{1}{3}$

12) $-\frac{4}{5}$

13) $\frac{1}{2}$

14) −1

15) $\frac{1}{3}$

16) $\frac{1}{8}$

17) $\frac{7}{5}$

18) 2

19) −3

20) 2

21) −1

22) 2

23) 2

24) −3

25) −1

26) $\frac{5}{2}$

27) 2

28) −11

Graphing Lines Using Line Equation

1) $y = x - 2$

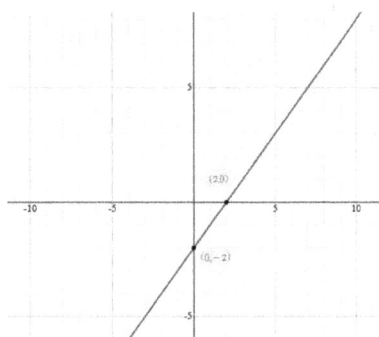

2) $y = -3x + 2$

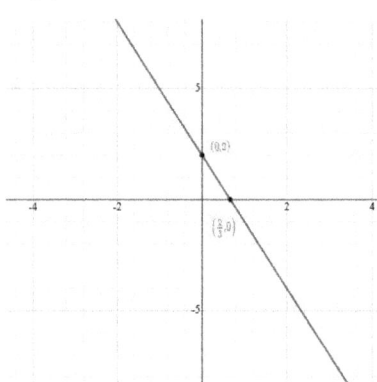

3) $x + y = 0$

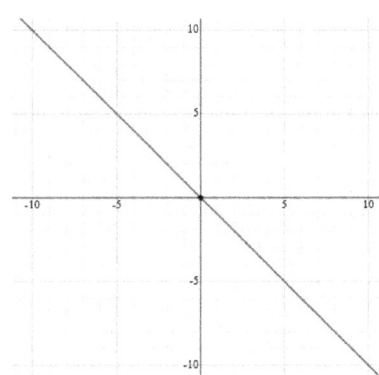

4) $x + y = -3$

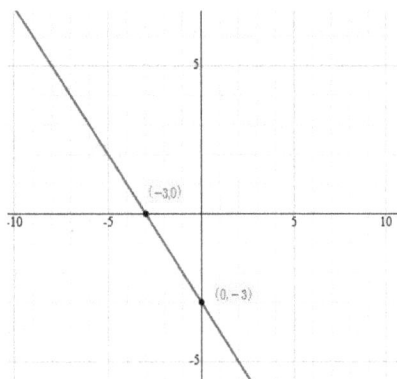

5) $2x + 3y = -4$

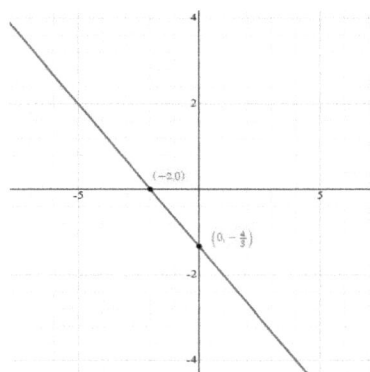

6) $y - 3x + 6 = 0$

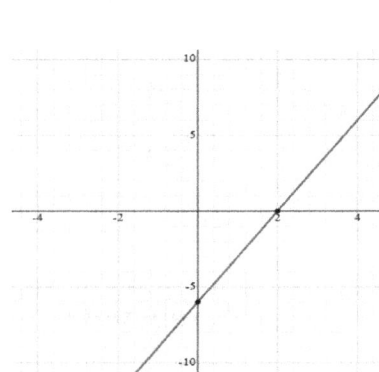

Writing Linear Equations

1) $y = 14x - 33$

2) $y = x + 9$

3) $y = -4x + 47$

4) $y = x - 4$

5) $y = 2x - 31$

6) $y = 9x - 43$

7) $y = -2x + 24$

8) $y = -3x + 1$

9) $y = -2x + 5$

10) $y = -2x + 6$

11) $y = 5x - 18$

12) $y = 14x - 122$

13) $y = 13x + 52$

14) $y = x + 10$

15) $y = -5x + 58$

16) $y = -5x + 5$

17) $y = 6x + 12$

18) $y = -11x - 4$

19) $y = -3x + 17$

20) $y = -5x - 11$

21) $y = -10x - 30$

22) $y = 8x + 7$

Graphing Linear Inequalities

1) $y > 4x - 5$

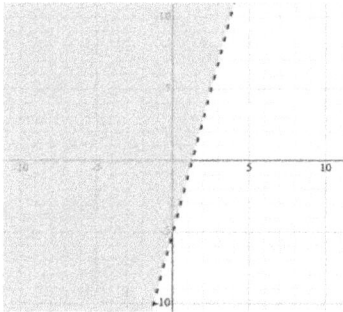

2) $y < 2x + 4$

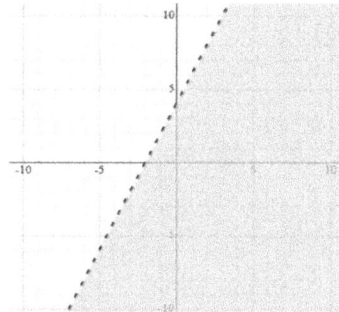

3) $y \leq -5x - 2$

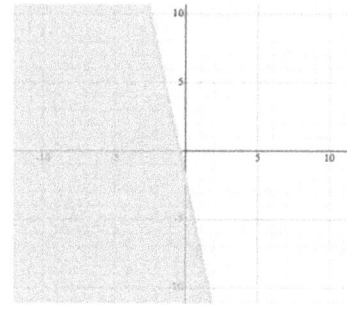

4) $4y \geq 12 + 4x$

5) $-12y < 3x - 24$

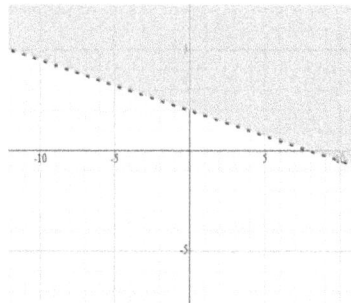

6) $5y \geq -15x + 10$

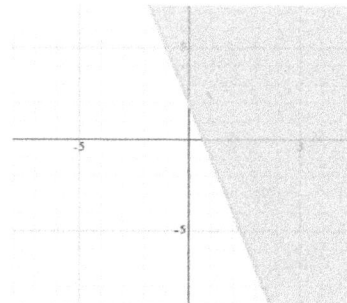

Finding Midpoint

1) $(-1, 0)$

2) $(4, 4)$

3) $(6, 4)$

4) $(-5, 1)$

5) $(8, -10)$

6) $(-3, -8)$

7) $(0, 10)$

8) $(-6, 4)$

9) $(6, -2)$

10) $(8, 6)$

11) $(-1, 6)$

12) $(4, 1)$

13) $(-4, 7.5)$

14) $(10, 2.5)$

15) $(2, 8)$

16) $(-4, 4)$

17) $(10, -10)$

18) $(4, -6)$

19) $(-1.5, -1)$

20) $(15, 5)$

21) $(4, -3)$

22) $(-9, 9)$

23) $(5, -4)$

24) $(16, 6)$

25) $(-9, 11)$

26) $(5, 3)$

27) $(14, 34)$

Finding Distance of Two Points

1) 15

2) 17

3) 2

4) 26

5) 9

6) 29

7) 25

8) 13

9) 11

10) 10

11) 10

12) 25

13) $4\sqrt{10}$

14) 20

15) 6

16) 25

17) 15

18) 10

19) 18

20) 12

21) 6 *square units*

22) 32 *units*

23) 20 *units*

Chapter 8 :

Polynomials

Topics that you will practice in this chapter:

- ✓ Writing Polynomials in Standard Form
- ✓ Simplifying Polynomials
- ✓ Adding and Subtracting Polynomials
- ✓ Multiplying Monomials
- ✓ Multiplying and Dividing Monomials
- ✓ Multiplying a Polynomial and a Monomial
- ✓ Multiplying Binomials
- ✓ Factoring Trinomials
- ✓ Operations with Polynomials

Mathematics is the supreme judge; from its decisions there is no appeal. – Tobias Dantzig

Writing Polynomials in Standard Form

✎ Write each polynomial in standard form.

1) $11x - 7x =$

2) $-5 + 19x - 19x =$

3) $6x^5 - 12x^3 =$

4) $12 + 17x^4 - 12 =$

5) $5x^2 + 4x - 9x^3 =$

6) $-3x^2 + 12x^5 =$

7) $5x + 8x^3 - 2x^8 =$

8) $-7x^3 + 4x - 9x^6 =$

9) $3x^2 + 22 - 6x =$

10) $3 - 4x + 9x^4 =$

11) $13x^2 + 28x - 8x^3 =$

12) $16 + 4x^2 - 2x^3 =$

13) $19x^2 - 9x + 9x^4 =$

14) $3x^4 - 7x^2 - 2x^3 =$

15) $-51 + 3x^2 - 8x^4 =$

16) $7x^2 - 8x^6 + 4x^4 - 15 =$

17) $6x^4 - 4x^5 + 16 - 3x^3 =$

18) $-2x^6 + 4x - 7x^2 - 5x =$

19) $11x^7 + 8x^5 - 5x^7 - 3x^2 =$

20) $2x^2 - 12x^5 + 8x^2 + 3x^6 =$

21) $4x^5 - 11x^7 - 6x^3 + 16x^5 =$

22) $6x^3 + 3x^5 + 34x^4 - 8x^5 =$

23) $3x(4x + 5 - 2x^2) =$

24) $12x(x^6 + 4x^3) =$

25) $5x(3x^2 + 6x + 4) =$

26) $7x(4 - 2x + 6x^5) =$

27) $3x(4x^4 - 4x^3 + 2) =$

28) $4x(2x^5 + 6x^2 - 3) =$

29) $5x(3x^4 + 4x^3 + 2x) =$

30) $2x(3x - 2x^3 + 4x^6) =$

Simplifying Polynomials

✎ **Simplify each expression.**

1) $3(4x - 20) =$

2) $5x(3x - 4) =$

3) $6x(5x - 7) =$

4) $3x(7x + 5) =$

5) $5x(4x - 3) =$

6) $6x(8x + 2) =$

7) $(3x - 2)(x - 4) =$

8) $(x - 5)(2x + 6) =$

9) $(x - 3)(x - 7) =$

10) $(3x + 4)(3x - 4) =$

11) $(5x - 4)(5x - 2) =$

12) $6x^2 + 6x^2 - 8x^4 =$

13) $3x - 2x^2 + 5x^3 + 7 =$

14) $7x + 4x^2 - 10x^3 =$

15) $12x^2 + 5x^5 - 6x^3 =$

16) $-5x^2 + 4x^6 + 6x^8 =$

17) $-12x^3 + 10x^5 - 4x^6 + 4x =$

18) $11 - 7x^2 + 4x^2 - 16x^3 + 11 =$

19) $2x^2 - 9x + 4x^3 + 15x - 10x =$

20) $13 - 7x^5 + 6x^5 - 4x^2 + 5 =$

21) $-5x^8 + x^6 - 14x^3 + 5x^8 =$

22) $(7x^4 - 4) + (7x^4 - 2x^4) =$

23) $3(3x^4 - 4x^3 - 6x^4) =$

24) $-5(x^9 + 8) - 5(10 - x^9) =$

25) $8x^3 - 9x^4 - 2x + 19 - 8x^3 =$

26) $11 - 8x^3 + 6x^3 - 7x^5 + 6 =$

27) $(5x^3 - 4x) - (6x - 2 - 6x^3) =$

28) $4x^2 - 5x^4 - x(3x^3 + 2x) =$

29) $6x + 6x^5 - 10 - 4(x^5 - 3) =$

30) $4 - 3x^4 + (6x^5 - 2x^4 + 5x^5) =$

31) $-(x^5 + 4) - 8(3 + x^5) =$

32) $(4x^3 - 3x) - (3x - 5x^3) =$

Adding and Subtracting Polynomials

✎ **Add or subtract expressions.**

1) $(-2x^2 - 3) + (3x^2 + 4) =$

11) $(15x^3 - 3x) - (3x - 4x^3) =$

2) $(4x^3 + 6) - (7 - 2x^3) =$

12) $(4x - x^5) - (6x^5 + 8x) =$

3) $(4x^5 + 5x^2) - (2x^5 + 15) =$

13) $(2x^2 - 7x^7) - (4x^7 - 6x) =$

4) $(6x^3 - 2x^2) + (5x^2 - 4x) =$

14) $(3x^2 - 5) + (8x^2 + 4x^5) =$

5) $(10x^4 + 28x) - (34x^4 + 6) =$

15) $(9x^4 + 5x^5) - (x^5 - 9x^4) =$

6) $(7x^2 - 3) + (7x^2 + 3) =$

16) $(-4x^3 - 2x) + (9x - 5x^3) =$

7) $(9x^2 + 4) - (10 - 5x^2) =$

17) $(4x - 3x^2) - (148x^2 + x) =$

8) $(6x^2 + x^5) - (x^5 + 4) =$

18) $(5x - 8x^4) - (3x^4 - 4x^2) =$

9) $(4x^3 - x) + (3x - 7x^3) =$

19) $(8x^4 - 4) + (2x^4 - 3x^2) =$

10) $(11x + 10) - (8x + 10) =$

20) $(5x^6 + 7x^3) - (x^3 - 5x^6) =$

21) $(-2x^2 + 20x^5 + 5x^4) + (12x^4 + 8x^5 + 24x^2) =$

22) $(7x^4 - 9x^7 - 6x) - (-3x^4 - 9x^7 + 6x) =$

23) $(14x + 12x^4 - 18x^6) + (20x^4 + 18x^6 - 10x) =$

24) $(5x^8 - 6x^6 - 4x) - (5x^3 + 9x^6 - 7x) =$

25) $(11x^2 - 6x^4 - 3x) - (-4x^2 - 12x^4 + 9x) =$

26) $(-5x^9 + 14x^3 + 3x^7) + (10x^7 + 26x^3 + 3x^9) =$

Multiplying Monomials

✎ **Simplify each expression.**

1) $6u^8 \times (-u^2) =$

2) $(-5p^8) \times (-2p^3) =$

3) $4xy^3z^5 \times 3z^4 =$

4) $3u^5t \times 8ut^4 =$

5) $(-5a^2) \times (-7a^3b^6) =$

6) $-3a^4b^3 \times 6a^2b =$

7) $13xy^5 \times x^4y^4 =$

8) $6p^4q^3 \times (-8pq^6) =$

9) $8s^4t^3 \times 4st^3 =$

10) $(-6x^4y^3) \times 6x^2y =$

11) $3xy^7z \times 12z^3 =$

12) $24xy \times x^2y =$

13) $13pq^4 \times (-3p^2q) =$

14) $13s^3t^4 \times st^4 =$

15) $11p^5 \times (-6p^3) =$

16) $(-8p^3q^5r) \times 3pq^4r^6 =$

17) $(-4a^4) \times (-7a^3b) =$

18) $6u^6v^2 \times (-5u^3v^4) =$

19) $9u^5 \times (-3u) =$

20) $-6xy^5 \times 4x^2y =$

21) $13y^5z^3 \times (-y^3z) =$

22) $8a^4bc^3 \times 2abc^3 =$

23) $(-7p^5q^6) \times (-5p^4q^2) =$

24) $4u^5v^3 \times (-4u^7v^3) =$

25) $17y^4z^5 \times (-y^6z) =$

26) $(-5pq^3r^2) \times 8p^2q^4r =$

27) $3ab^5c^6 \times 5a^4bc^2 =$

28) $6x^3yz^2 \times 3x^2y^7z^3 =$

Multiplying and Dividing Monomials

✍ **Simplify each expression.**

1) $(5x^5)(2x^2) =$

2) $(4x^4)(6x^2) =$

3) $(3x^4)(7x^4) =$

4) $(5x^6)(4x^2) =$

5) $(12x^4)(3x^6) =$

6) $(4yx^8)(8y^4x^3) =$

7) $(14x^4y)(x^3y^5) =$

8) $(-5x^3y^4)(2x^3y^5) =$

9) $(-6x^4y^2)(-3x^3y^5) =$

10) $(5x^3y)(-5x^2y^3) =$

11) $(6x^4y^3)(4x^3y^4) =$

12) $(4x^3y^2)(5x^2y^4) =$

13) $(12x^3y^6)(4x^4y^{10}) =$

14) $(15x^3y^5)(3x^4y^6) =$

15) $(7x^2y^7)(8x^6y^7) =$

16) $(-3x^3y^8)(7x^9y^4) =$

17) $\dfrac{5x^6y^6}{xy^4} =$

18) $\dfrac{19x^7y^5}{19x^6y} =$

19) $\dfrac{56x^4y^4}{8xy} =$

20) $\dfrac{81x^5y^6}{9x^4y^5} =$

21) $\dfrac{36x^7y^6}{9x^2y^3} =$

22) $\dfrac{48x^9y^7}{4x^4y^6} =$

23) $\dfrac{88x^{18}y^{12}}{11x^8y^9} =$

24) $\dfrac{30x^7y^6}{6x^8y^3} =$

25) $\dfrac{150x^7y^6}{30x^4y^6} =$

26) $\dfrac{-42x^{18}y^{14}}{6x^4y^9} =$

27) $\dfrac{-36x^7y^8}{9x^5y^8} =$

Multiplying a Polynomial and a Monomial

✍ **Find each product.**

1) $x(2x + 4) =$

2) $6(4 - 2x) =$

3) $5x(4x + 2) =$

4) $x(-4x + 5) =$

5) $8x(2x - 2) =$

6) $6(2x - 4y) =$

7) $7x(5x - 5) =$

8) $3x(12x + 2y) =$

9) $4x(x + 6y) =$

10) $11x(3x + 4y) =$

11) $7x(3x + 2) =$

12) $10x(4x - 10y) =$

13) $9x(3x - 2y) =$

14) $7x(x - 4y + 6) =$

15) $8x(2x^2 + 5y^2) =$

16) $12x(2x + 3y) =$

17) $4(2x^4 - 4y^4) =$

18) $4x(-3x^2y + 4y) =$

19) $-4(5x^3 - 2xy + 4) =$

20) $4(x^2 - 5xy - 6) =$

21) $8x(2x^3 - 5xy + 2x) =$

22) $-6x(-2x^3 - 6x + 2xy) =$

23) $3(2x^2 + xy - 9y^2) =$

24) $4x(5x^3 - 3x + 7) =$

25) $6(3x^{22} - 2x - 5) =$

26) $x^2(-2x^3 + 4x + 3) =$

27) $x^2(4x^3 + 10 - 2x) =$

28) $4x^4(3x^3 - 2x + 5) =$

29) $2x^2(4x^4 - 5xy + 7y^3) =$

30) $5x^2(5x^4 - 3x + 9) =$

31) $7x^2(6x^2 + 3x - 6) =$

32) $4x(x^3 - 4xy + 2y^2) =$

Multiplying Binomials

✍ **Find each product.**

1) $(x + 3)(x + 6) =$

2) $(x - 4)(x + 3) =$

3) $(x - 3)(x - 8) =$

4) $(x + 8)(x + 9) =$

5) $(x - 2)(x - 12) =$

6) $(x + 5)(x + 5) =$

7) $(x - 6)(x + 7) =$

8) $(x - 8)(x - 3) =$

9) $(x + 7)(x + 12) =$

10) $(x - 4)(x + 8) =$

11) $(x + 8)(x + 8) =$

12) $(x + 2)(x + 7) =$

13) $(x - 6)(x + 6) =$

14) $(x - 5)(x + 5) =$

15) $(x + 11)(x + 11) =$

16) $(x + 6)(x + 9) =$

17) $(x - 2)(x + 2) =$

18) $(x - 4)(x + 7) =$

19) $(3x + 5)(x + 6) =$

20) $(5x - 6)(4x + 8) =$

21) $(x - 7)(3x + 7) =$

22) $(x - 9)(x - 4) =$

23) $(x - 12)(x + 2) =$

24) $(2x - 4)(5x + 4) =$

25) $(3x - 8)(x + 8) =$

26) $(7x - 2)(6x + 3) =$

27) $(4x + 5)(3x + 5) =$

28) $(7x - 4)(9x + 4) =$

29) $(x + 2)(2x - 8) =$

30) $(5x - 4)(5x + 4) =$

31) $(3x + 2)(3x - 7) =$

32) $(x^2 + 8)(x^2 - 8) =$

Factoring Trinomials

✎ **Factor each trinomial.**

1) $x^2 + 8x + 12 =$

2) $x^2 - 6x + 5 =$

3) $x^2 + 15x + 36 =$

4) $x^2 - 12x + 35 =$

5) $x^2 - 11x + 18 =$

6) $x^2 - 9x + 18 =$

7) $x^2 + 18x + 72 =$

8) $x^2 - x - 72 =$

9) $x^2 + 4x - 21 =$

10) $x^2 - 13x + 22 =$

11) $x^2 + 2x - 24 =$

12) $x^2 - 3x - 40 =$

13) $x^2 - 3x - 70 =$

14) $x^2 + 26x + 169 =$

15) $4x^2 - 7x - 15 =$

16) $x^2 - 14x + 33 =$

17) $10x^2 + 5x - 15 =$

18) $6x^2 - 4x - 42 =$

19) $x^2 + 12x + 36 =$

20) $5x^2 + 17x - 12 =$

✎ **Calculate each problem.**

21) The area of a rectangle is $x^2 - x - 56$. If the width of rectangle is $x + 7$, what is its length? _____

22) The area of a parallelogram is $4x^2 + 17x - 15$ and its height is $x + 5$. What is the base of the parallelogram? _____

23) The area of a rectangle is $6x^2 - 22x + 12$. If the width of the rectangle is $3x - 2$, what is its length? _____

Operations with Polynomials

✎ **Find each product.**

1) $4(5x + 3) =$ _____

2) $8(2x + 6) =$ _____

3) $2(5x - 2) =$ _____

4) $-4(7x - 3) =$ _____

5) $3x^2(9x + 1) =$ _____

6) $4x^6(7x - 9) =$ _____

7) $3x^4(-7x + 3) =$ _____

8) $-8x^4(5x - 8) =$ _____

9) $7(x^2 + 5x - 3) =$ _____

10) $9(5x^2 - 7x + 5) =$ _____

11) $3(3x^2 + 3x + 2) =$ _____

12) $5x(3x^2 + 5x + 8) =$ _____

13) $(5x + 7)(3x - 3) =$ _____

14) $(9x + 3)(3x - 5) =$ _____

15) $(6x + 3)(4x - 2) =$ _____

16) $(7x - 2)(3x + 5) =$ _____

✎ **Calculate each problem.**

17) The measures of two sides of a triangle are $(2x + 5y)$ and $(6x - 3y)$. If the perimeter of the triangle is $(13x + 4y)$, what is the measure of the third side? _____

18) The height of a triangle is $(8x + 5)$ and its base is $(4x - 3)$. What is the area of the triangle? _____

19) One side of a square is $(6x + 2)$. What is the area of the square? _____

20) The length of a rectangle is $(5x - 8y)$ and its width is $(15x + 8y)$. What is the perimeter of the rectangle? _____

21) The side of a cube measures $(x + 2)$. What is the volume of the cube? _____

22) If the perimeter of a rectangle is $(28x + 6y)$ and its width is $(5x + 2y)$, what is the length of the rectangle? _____

Answers of Worksheets

Writing Polynomials in Standard Form

1) $4x$

2) -5

3) $6x^5 - 12x^3$

4) $14x^4$

5) $-9x^3 + 5x^2 + 4x$

6) $12x^5 - 3x^2$

7) $-2x^8 + 8x^3 + 5x$

8) $-9x^6 - 7x^3 + 4x$

9) $3x^2 - 6x + 22$

10) $9x^4 - 4x + 3$

11) $-8x^3 + 13x^2 + 28x$

12) $-2x^3 + 4x^2 + 16$

13) $9x^4 + 19x^2 - 9x$

14) $3x^4 - 2x^3 - 7x^2$

15) $-8x^4 + 3x^2 - 51$

16) $-8x^6 + 4x^4 + 7x^2 - 15$

17) $-4x^5 + 6x^4 - 3x^3 + 16$

18) $-2x^6 - 7x^2 - x$

19) $6x^7 + 8x^5 - 3x^2$

20) $3x^6 - 12x^5 + 10x^2$

21) $-11x^7 + 20x^5 - 6x^3$

22) $-5x^5 + 34x^4 + 6x^3$

23) $-6x^3 + 12x^2 + 15x$

24) $12x^7 + 48x^4$

25) $15x^3 + 30x^2 + 20x$

26) $42x^6 - 14x^2 + 28x$

27) $12x^5 - 12x^4 + 6x$

28) $8x^6 + 24x^3 - 12x$

29) $15x^5 + 20x^4 + 10x^2$

30) $8x^7 - 4x^4 + 6x^2$

Simplifying Polynomials

1) $12x - 60$

2) $15x^2 - 20x$

3) $30x^2 - 42x$

4) $21x^2 + 15x$

5) $20x^2 - 15x$

6) $48x^2 + 12x$

7) $3x^2 - 14x + 8$

8) $2x^2 - 4x - 30$

9) $x^2 - 10x + 21$

10) $9x^2 - 16$

11) $25x^2 - 30x + 8$

12) $-8x^4 + 12x^2$

13) $5x^3 - 2x^2 + 3x + 7$

14) $-10x^3 + 4x^2 + 7x$

15) $5x^5 - 6x^3 + 12x^2$

16) $6x^8 + 4x^6 - 5x^2$

17) $-4x^6 + 10x^5 - 12x^3 + 4x$

18) $-16x^3 - 3x^2 + 22$

19) $4x^3 + 2x^2 - 4x$

20) $-x^5 - 4x^2 + 18$

21) $x^6 - 14x^3$

22) $12x^4 - 4$

23) $-9x^4 - 12x^3$

24) -90

25) $-9x^4 - 2x + 19$

26) $-7x^5 - 2x^3 + 17$

27) $11x^3 - 10x + 2$

28) $-8x^4 + 2x^2$

29) $2x^5 + 6x + 2$

30) $11x^5 - 5x^4 + 4$

31) $-9x^5 - 28$

32) $9x^3 - 6x$

Adding and Subtracting Polynomials

1) $x^2 + 1$

2) $6x^3 - 1$

3) $2x^5 + 5x^2 - 15$

4) $6x^3 + 3x^2 - 4x$

5) $-24x^4 + 28x - 6$

6) $14x^2$

7) $14x^2 - 6$

8) $6x^2 - 4$

9) $-3x^3 + 2x$

10) $3x$

11) $19x^3 - 6x$

12) $-7x^5 - 4x$

13) $-11x^7 + 2x^2 + 6x$

14) $4x^5 + 11x^2 - 5$

15) $4x^5 + 18x^4$

16) $-9x^3 + 7x$

17) $-151x^2 + 3x$

18) $-11x^4 + 4x^2 + 5x$

19) $10x^4 - 3x^2 - 4$

20) $10x^6 + 6x^3$

21) $28x^5 + 17x^4 + 22x^2$

22) $10x^4 - 12x$

23) $32x^4 + 4x$

24) $5x^8 - 15x^6 - 5x^3 + 3x$

25) $6x^4 + 15x^2 - 12x$

26) $-2x^9 + 13x^7 + 40x^3$

Multiplying Monomials

1) $-6u^{10}$

2) $10p^{11}$

3) $12xy^3z^9$

4) $24u^6t^5$

5) $35a^5b^6$

6) $-18a^6b^4$

7) $13x^5y^9$

8) $-48p^5q^9$

9) $32s^5t^6$

10) $-36x^6y^4$

11) $36xy^7z^4$

12) $24px^3y^2$

13) $-39p^3q^5$

14) $13s^4t^8$

15) $-66p^8$

16) $-24p^4q^9r^7$

17) $28a^7b$

18) $-30u^9v^6$

19) $-27u^6$

20) $-24x^3y^6$

21) $-13y^8z^4$

22) $16a^5b^2c^6$

23) $35p^9q^8$

24) $-16u^{12}v^6$

25) $-17y^{10}z^6$

26) $-40p^3q^7r^3$

27) $15a^5b^6c^8$

28) $18x^5y^8z^5$

Multiplying and Dividing Monomials

1) $10x^7$

2) $24x^6$

3) $21x^8$

4) $20x^8$

5) $36x^{10}$

6) $32x^{11}y^5$

7) $14x^7y^6$

8) $-10x^6y^9$

9) $18x^7y^7$

10) $-25x^5y^4$

11) $24x^7y^7$

12) $20x^5y^6$

13) $48x^7y^{16}$

14) $45x^7y^{11}$

15) $56x^8y^{14}$

16) $-21x^{12}y^{12}$

17) $5x^5y^2$

18) xy^4

19) $7x^3y^3$

20) $9xy$

21) $4x^5y^3$

22) $12x^5y$

23) $8x^{10}y^3$

24) $5x^{-1}y^3$

25) $5x^3$

26) $-7x^{14}y^5$

27) $-4x^2$

Multiplying a Polynomial and a Monomial

1) $2x^2 + 4x$

2) $-12x + 24$

3) $20x^2 + 10x$

4) $-4x^2 + 5x$

5) $16x^2 - 16x$

6) $12x - 24y$

7) $35x^2 - 35x$

8) $36x^2 + 6xy$

9) $4x^2 + 24xy$

10) $33x^2 + 44xy$

11) $21x^2 + 14x$

12) $40x^2 - 100xy$

13) $27x^2 - 18xy$

14) $7x^2 - 28xy + 42x$

15) $16x^3 + 40xy^2$

16) $24x^2 + 36xy$

17) $8x^4 - 16y^4$

18) $-12x^3y + 16xy$

19) $-20x^3 + 8xy - 16$

20) $4x^2 - 20xy - 24$

21) $16x^4 - 40x^2y + 16x^2$

22) $12x^4 + 36x^2 - 12x^2y$

23) $6x^2 + 3xy - 27y^2$

24) $20x^4 - 12x^2 + 28x$

25) $18x^{22} - 12x - 30$

26) $-2x^5 + 4x^3 + 3x^2$

27) $4x^5 - 2x^3 + 10x^2$

28) $12x^7 - 8x^5 + 20x^4$

29) $8x^6 - 10x^3y + 14x^2y^3$

30) $25x^6 - 15x^3 + 45x^2$

31) $42x^4 + 21x^3 - 42x^2$

32) $4x^4 - 16x^2y + 8xy^2$

Multiplying Binomials

1) $x^2 + 9x + 18$

2) $x^2 - x - 12$

3) $x^2 - 11x + 24$

4) $x^2 + 17x + 72$

5) $x^2 - 14x + 24$

6) $x^2 + 10x + 25$

7) $x^2 + x - 42$

8) $x^2 - 11x + 24$

9) $x^2 + 19x + 84$

10) $x^2 + 4x - 32$

11) $x^2 + 16x + 64$

12) $x^2 + 9x + 14$

13) $x^2 - 36$

14) $x^2 - 25$

15) $x^2 + 22x + 121$

16) $x^2 + 15x + 54$

17) $x^2 - 4$

18) $x^2 + 3x - 28$

19) $3x^2 + 23x + 30$

20) $20x^2 + 16x - 48$

21) $3x^2 - 14x - 49$

22) $x^2 - 13x + 36$

23) $x^2 - 10x - 24$

24) $10x^2 - 12x - 16$

25) $3x^2 + 16x - 64$

26) $42x^2 + 9x - 6$

27) $12x^2 + 35x + 25$

28) $63x^2 - 8x - 16$

29) $2x^2 - 4x - 16$

30) $25x^2 - 16$

31) $9x^2 - 15x - 14$

32) $x^4 - 64$

Factoring Trinomials

1) $(x + 6)(x + 2)$

2) $(x - 5)(x - 1)$

3) $(x + 12)(x + 3)$

4) $(x - 5)(x - 7)$

5) $(x - 2)(x - 9)$

6) $(x - 6)(x - 3)$

7) $(x + 6)(x + 12)$

8) $(x + 8)(x - 9)$

9) $(x - 3)(x + 7)$

10) $(x - 11)(x - 2)$

11) $(x - 4)(x + 6)$

12) $(x - 8)(x + 5)$

13) $(x + 7)(x - 10)$

14) $(x + 13)(x + 13)$

15) $(4x + 5)(x - 3)$

16) $(x - 11)(x - 3)$

17) $(5x - 5)(2x + 3)$

18) $(2x - 6)(3x + 7)$

19) $(x + 6)(x + 6)$

20) $(5x - 3)(x + 4)$

21) $(x - 8)$

22) $(4x - 3)$

23) $(2x - 6)$

Operations with Polynomials

1) $20x + 12$

2) $16x + 48$

3) $10x - 4$

4) $-28x + 12$

5) $27x^3 + 3x^2$

6) $28x^7 - 36x^6$

7) $-21x^5 + 9x^4$

8) $-40x^5 + 64x^4$

9) $7x^2 + 35x - 21$

10) $45x^2 - 63x + 45$

11) $9x^2 + 9x + 6$

12) $15x^3 + 25x^2 + 40x$

13) $15x^2 + 6x - 21$

14) $27x^2 - 36x - 15$

15) $24x^2 - 6$

16) $21x^2 + 29x - 10$

17) $(5x + 2y)$

18) $16x^2 - 2x - \frac{15}{2}$

19) $36x^2 + 24x + 4$

20) $40x$

21) $x^3 + 6x^2 + 12x + 8$

22) $(9x + y)$

Chapter 9 :
Functions Operations and Quadratic

Topics that you will practice in this chapter:

- ✓ Evaluating Function
- ✓ Adding and Subtracting Functions
- ✓ Multiplying and Dividing Functions
- ✓ Composition of Functions
- ✓ Quadratic Equation
- ✓ Solving Quadratic Equations
- ✓ Quadratic Formula and the Discriminant
- ✓ Quadratic Inequalities
- ✓ Graphing Quadratic Functions
- ✓ Domain and Range of Radical Functions
- ✓ Solving Radical Equations

It's fine to work on any problem, so long as it generates interesting mathematics along the way – even if you don't solve it at the end of the day." – Andrew Wiles

Evaluating Function

✍ **Write each of following in function notation.**

1) $h = -8x + 3$

2) $k = 2a - 14$

3) $d = 11t$

4) $y = \frac{5}{12}x - \frac{7}{12}$

5) $m = 24n - 210$

6) $c = p^2 - 5p + 10$

✍ **Evaluate each function.**

7) $f(x) = 2x - 7$, find $f(-3)$

8) $g(x) = \frac{1}{9}x + 12$, find $f(18)$

9) $h(x) = -4x + 9$, find $f(3)$

10) $f(x) = -x + 19$, find $f(-3)$

11) $f(a) = 7a - 12$, find $f(3)$

12) $h(x) = 14 - 3x$, find $f(-4)$

13) $g(n) = 6n - 10$, find $f(2)$

14) $f(x) = -11x - 4$, find $f(-1)$

15) $k(n) = -20 - 3.5n$, find $f(2)$

16) $f(x) = -0.7x + 3.3$, find $f(-7)$

17) $g(n) = \frac{11n+8}{n}$, find $g(2)$

18) $g(n) = \sqrt{3n} + 12$, find $g(3)$

19) $h(x) = x^{-2} - 7$, find $h(\frac{1}{9})$

20) $h(n) = n^{-3} + 11$, find $h(\frac{1}{4})$

21) $h(n) = n^3 - 2$, find $h(\frac{1}{2})$

22) $h(n) = n^2 - 4$, find $h(-\frac{1}{3})$

23) $h(n) = 4n^2 - 13$, find $h(-5)$

24) $h(n) = -2n^3 - 6n$, find $h(2)$

25) $g(n) = \sqrt{16n^2} - \sqrt{n}$, find $g(4)$

26) $h(a) = \frac{-14a+9}{3a}$, find $h(-b)$

27) $k(a) = 12a - 14$, find $k(a - 3)$

28) $h(x) = \frac{1}{9}x + 18$, find $h(-18x)$

29) $h(x) = 8x^2 + 16$, find $h(\frac{x}{2})$

30) $h(x) = x^4 - 20$, find $h(-2x)$

Adding and Subtracting Functions

✎ **Perform the indicated operation.**

1) $f(x) = 2x + 3$

   $g(x) = x + 7$

   Find $(f - g)(2)$

2) $g(a) = -5a - 8$

   $f(a) = -3a - 5$

   Find $(g - f)(-2)$

3) $h(t) = 4t + 3$

   $g(t) = 4t + 7$

   Find $(h - g)(t)$

4) $g(a) = -6a - 10$

   $f(a) = 3a^2 + 9$

   Find $(g - f)(x)$

5) $g(x) = \frac{5}{6}x - 23$

   $h(x) = \frac{5}{12}x + 25$

   Find $g(12) - h(12)$

6) $h(x) = \sqrt{3x} - 2$

   $g(x) = \sqrt{3x} + 5$

   Find $(h + g)(12)$

7) $f(x) = x^{-1}$

   $g(x) = x^2 + \frac{5}{x}$

   Find $(f - g)(-3)$

8) $h(n) = n^2 + 2$

   $g(n) = -4n + 6$

   Find $(h - g)(2a)$

9) $g(x) = -2x^2 - 5 - 4x$

   $f(x) = 7 + 2x$

   Find $(g - f)(3x)$

10) $g(t) = 11t - 4$

   $f(t) = -2t^2 + 5$

   Find $(g + f)(-t)$

11) $f(x) = 8x + 9$

   $g(x) = -5x^2 + 3x$

   Find $(f - g)(-x^2)$

12) $f(x) = -3x^4 - 5x$

   $g(x) = 2x^4 + 5x + 22$

   Find $(f + g)(3x^2)$

Multiplying and Dividing Functions

✎ **Perform the indicated operation.**

1) $g(x) = -2x - 1$

$f(x) = 4x + 3$

Find $(g.f)(2)$

2) $f(x) = 5x$

$h(x) = -2x + 3$

Find $(f.h)(-2)$

3) $g(a) = 5a - 2$

$h(a) = 2a - 3$

Find $(g.h)(-3)$

4) $f(x) = 2x - 7$

$h(x) = x - 5$

Find $(\frac{f}{h})(4)$

5) $f(x) = 8a^2$

$g(x) = 3 + 2a$

Find $(\frac{f}{g})(2)$

6) $g(a) = \sqrt{4a} + 2$

$f(a) = (-a)^4 + 1$

Find $(\frac{g}{f})(1)$

7) $g(t) = t^3 + 1$

$h(t) = 5t - 2$

Find $(g.h)(-2)$

8) $g(n) = n^2 + 2n - 4$

$h(n) = -5n + 3$

Find $(g.h)(1)$

9) $g(a) = (a - 3)^2$

$f(a) = a^2 + 4$

Find $(\frac{g}{f})(3)$

10) $g(x) = -3x^2 + \frac{4}{5}x + 9$

$f(x) = x^2 - 24$

Find $(\frac{g}{f})(5)$

11) $f(x) = 2x^3 - 5x^2 + 1$

$g(x) = 3x - 1$

Find $(f.g)(x)$

12) $f(x) = 5x - 2$

$g(x) = x^3 - 2x$

Find $(f.g)(x^2)$

Composition of Functions

✎ Using $f(x) = 2x - 5$ and $g(x) = -2x$, find:

1) $f(g(2)) =$

2) $f(g(-1)) =$

3) $g(f(-4)) =$

4) $g(f(5)) =$

5) $f(g(3)) =$

6) $g(f(0)) =$

✎ Using $f(x) = -\frac{1}{4}x + \frac{3}{4}$ and $g(x) = 2x^2$, find:

7) $g(f(-2)) =$

8) $g(f(4)) =$

9) $g(g(1)) =$

10) $f(f(1)) =$

11) $g(f(-4)) =$

12) $f(g(x)) =$

✎ Using $f(x) = -2x + 2$ and $g(x) = x + 1$, find:

13) $g(f(1)) =$

14) $f(f(0)) =$

15) $f(g(-1)) =$

16) $f(g(-3)) =$

17) $g(f(2)) =$

18) $f(g(x)) =$

✎ Using $f(x) = \sqrt{x + 9}$ and $g(x) = x - 9$, find:

19) $f(g(9)) =$

20) $g(f(-9)) =$

21) $f(g(4)) =$

22) $f(f(7)) =$

23) $g(f(-5)) =$

24) $g(g(0)) =$

Quadratic Equation

✎ Multiply.

1) $(x - 4)(x + 6) = $ _____

2) $(x + 5)(x + 7) = $ _____

3) $(x - 6)(x + 8) = $ _____

4) $(x + 2)(x - 9) = $ _____

5) $(x - 7)(x - 8) = $ _____

6) $(3x + 2)(x - 3) = $ _____

7) $(4x - 3)(x + 2) = $ _____

8) $(4x - 5)(x + 1) = $ _____

9) $(7x + 1)(x - 6) = $ _____

10) $(5x + 1)(3x - 3) = $ _____

✎ Factor each expression.

11) $x^2 - 2x - 8 = $ _____

12) $x^2 + 8x + 15 = $ _____

13) $x^2 - 2x - 24 = $ _____

14) $x^2 - 10x + 21 = $ _____

15) $x^2 + 10x + 21 = $ _____

16) $4x^2 + 9x + 5 = $ _____

17) $5x^2 + 13x - 6 = $ _____

18) $5x^2 + 17x - 12 = $ _____

19) $2x^2 + 7x + 5 = $ _____

20) $9x^2 - 21x + 6 = $ _____

✎ Calculate each equation.

21) $(x + 6)(x - 3) = 0$

22) $(x + 1)(x + 8) = 0$

23) $(3x + 6)(x + 5) = 0$

24) $(2x - 2)(4x + 8) = 0$

25) $x^2 + x + 10 = 22$

26) $x^2 + 11x + 36 = 12$

27) $2x^2 + 9x + 9 = 5$

28) $x^2 + 3x - 24 = 4$

29) $5x^2 + 5x - 40 = 20$

30) $8x^2 + 8x = 48$

Solving Quadratic Equations

✎ **Solve each equation by factoring or using the quadratic formula.**

1) $(x + 9)(x - 1) = 0$

2) $(x + 7)(x + 6) = 0$

3) $(x - 8)(x + 3) = 0$

4) $(x - 6)(x - 4) = 0$

5) $(x + 2)(x + 12) = 0$

6) $(5x + 4)(x + 7) = 0$

7) $(6x + 1)(4x + 5) = 0$

8) $(2x + 7)(x + 8) = 0$

9) $(x + 6)(3x + 15) = 0$

10) $(12x + 2)(x + 8) = 0$

11) $x^2 = 8x$

12) $x^2 - 16 = 0$

13) $3x^2 + 6 = 9x$

14) $-2x^2 - 8 = 10x$

15) $5x^2 + 40x = 45$

16) $x^2 + 10x = 24$

17) $x^2 + 6x = 16$

18) $x^2 + 9x = -18$

19) $x^2 + 13x = -36$

20) $x^2 + 3x - 15 = 5x$

21) $x^2 + 8x + 7 = -8$

22) $3x^2 - 11x = -9 + x$

23) $10x^2 + 3 = 27x - 15$

24) $7x^2 - 6x + 8 = 8$

25) $2x^2 - 12 = -3x + 2$

26) $10x^2 - 26x - 3 = -15$

27) $3x^2 + 21 = -16x + 5$

28) $x^2 + 15x - 10 = -66$

29) $3x^2 - 8x - 8 = 4 + x$

30) $2x^2 + 6x - 24 = 12$

31) $3x^2 - 33x + 54 = -18$

32) $-10x^2 - 15x - 9 = -9 - 27x^2$

Quadratic Formula and the Discriminant

✍ **Find the value of the discriminant of each quadratic equation.**

1) $3x(x - 8) = 0$

2) $2x^2 + 6x - 4 = 0$

3) $x^2 + 6x + 7 = 0$

4) $x^2 - x + 3 = 0$

5) $x^2 + 4x - 3 = 0$

6) $2x^2 + 6x - 10 = 0$

7) $3x^2 + 7x + 5 = 0$

8) $x^2 - 6x - 4 = 0$

9) $2x^2 + 8x + 3 = 0$

10) $x^2 + 7x - 5 = 0$

11) $5x^2 + 2x - 3 = 0$

12) $-3x^2 - 11x + 4 = 0$

13) $-6x^2 - 12x + 8 = 0$

14) $-x^2 - 9x - 12 = 0$

15) $7x^2 - 6x - 10 = 0$

16) $-4x^2 - 2x + 8 = 0$

17) $5x^2 + 8x - 2 = 0$

18) $6x^2 - 4x = 0$

19) $3x^2 - 5x + 2 = 0$

20) $4x^2 + 9x + 3 = 0$

✍ **Find the discriminant of each quadratic equation then state the number of real and imaginary solutions.**

21) $-4x^2 - 16 = 16x$

22) $20x^2 = 20x - 5$

23) $-11x^2 - 19x = 26$

24) $22x^2 - 4x + 1 = 18x^2$

25) $-11x^2 = -15x + 8$

26) $3x^2 + 6x + 9 = 6$

27) $13x^2 - 5x - 12 = -26$

28) $-8x^2 - 32x - 25 = 7$

Graphing Quadratic Functions

✎ketch the graph of each function. Identify the vertex and axis of symmetry.

1) $y = (x + 3)^2 + 2$

2) $y = (x - 3)^2 - 2$

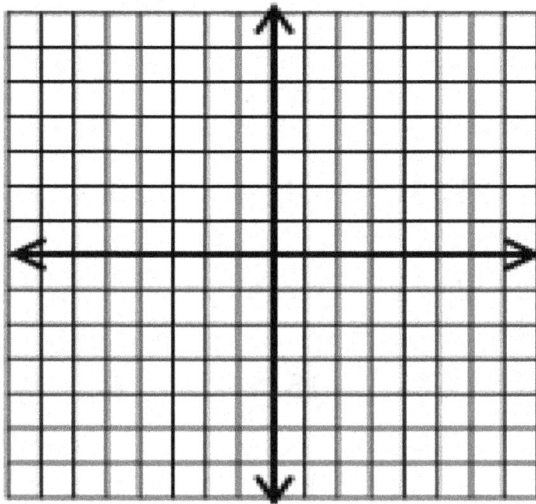

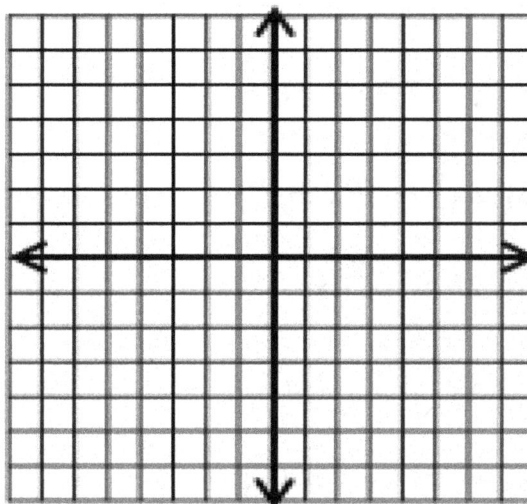

3) $y = 6 - (-x + 4)^2$

4) $y = -3x^2 - 6x + 9$

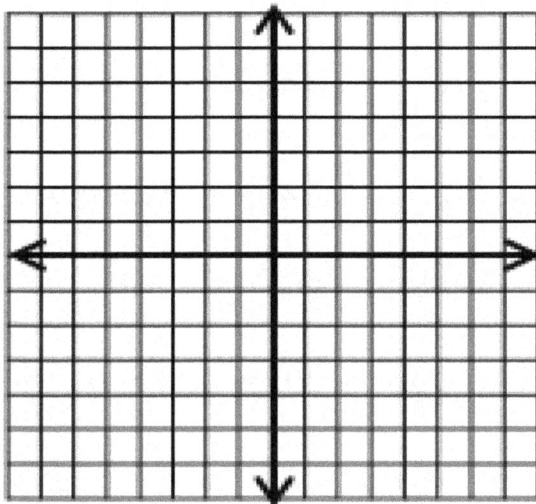

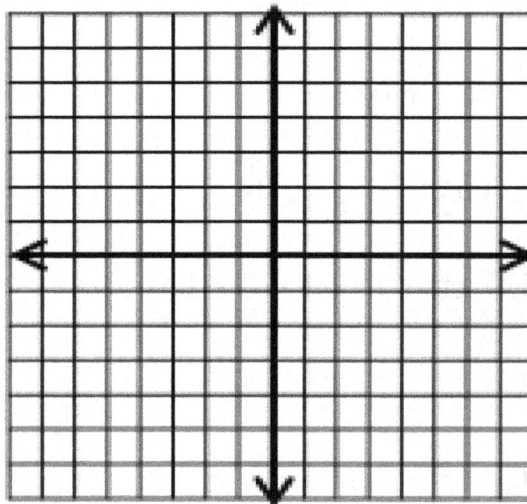

Quadratic Inequalities

✏ **Solve each quadratic inequality.**

1) $x^2 - 25 < 0$

2) $-x^2 - 6x - 8 > 0$

3) $5x^2 + 15x + 30 < 0$

4) $x^2 + 8x + 16 > 0$

5) $2x^2 - 18x - 20 \geq 0$

6) $x^2 > -10x - 25$

7) $3x^2 + 2x + 16 \leq 0$

8) $x^2 - 5x - 14 \leq 0$

9) $x^2 - 6x - 7 \geq 0$

10) $2x^2 + 16x - 18 < 0$

11) $x^2 + 6x - 72 > 0$

12) $3x^2 - 3x - 36 > 0$

13) $x^2 - 15x + 64 \leq 0$

14) $2x^2 - 24x + 72 \leq 0$

15) $x^2 - 16x + 63 \geq 0$

16) $x^2 - 16x + 55 \geq 0$

17) $x^2 - 81 \leq 0$

18) $x^2 - 17x + 42 \geq 0$

19) $9x^2 + 14x + 36 \leq 0$

20) $4x^2 - 2x - 24 > 2x^2$

21) $5x^2 - 20x + 20 < 0$

22) $7x^2 - 6x \geq 6x^2 - 5$

23) $5x^2 - 15 > 4x^2 + 2x$

24) $3x^2 - 4x \geq 3x^2 - 9x + 15$

25) $8x^2 + 9x - 54 > 5x^2$

26) $10x^2 + 50x - 60 < 0$

27) $-x^2 + 15x - 57 \geq 0$

28) $-5x^2 + 25x + 30 \leq 0$

29) $5x^2 + 40x + 75 < 0$

30) $9x^2 + 20x + 180 \leq 0$

31) $3x^2 + 2x - 36 \geq -x$

32) $3x^2 + 9x + 9 \leq 6x^2 + 3x$

Domain and Range of Radical Functions

✍ **Identify the domain and range of each function.**

1) $y = \sqrt{x + 8} - 7$

2) $y = \sqrt[3]{3x - 5} - 4$

3) $y = \sqrt{3x - 9} + 3$

4) $y = \sqrt[3]{(4x + 6)} - 2$

5) $y = 3\sqrt{4x + 20} + 6$

6) $y = \sqrt[3]{(5x - 2)} - 11$

7) $y = 4\sqrt{9x^2 + 8} + 3$

8) $y = \sqrt[3]{(7x^2 - 2)} - 6$

9) $y = 2\sqrt{2x^3 + 16} - 3$

10) $y = \sqrt[3]{(11x + 4)} - 2x$

11) $y = 3\sqrt{-2(4x + 8)} + 5$

12) $y = \sqrt[5]{(3x^2 - 12)} - 6$

13) $y = 3\sqrt{x - 5} - 2$

14) $y = \sqrt[3]{6x + 9} - 4$

✍ **Sketch the graph of each function.**

15) $y = -3\sqrt{x} + 5$

16) $y = 3\sqrt{x} - 6$

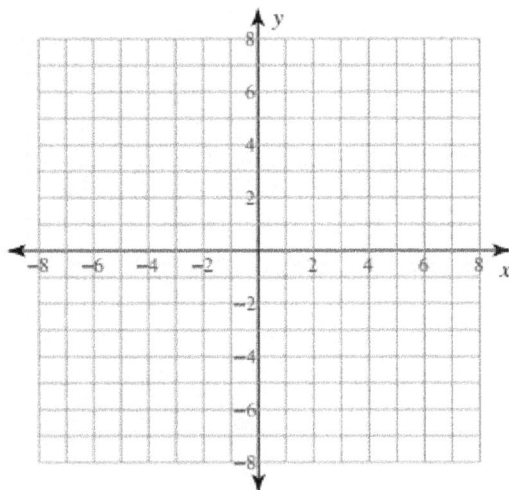

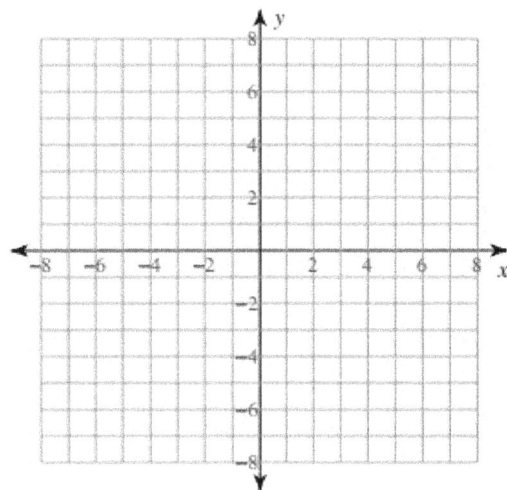

Solving Radical Equations

✍ **Solve each equation. Remember to check for extraneous solutions.**

1) $\sqrt{a} = 9$

2) $\sqrt{v} = 6$

3) $\sqrt{r} = 4$

4) $8 = 16\sqrt{x}$

5) $\sqrt{x + 3} = 18$

6) $6 = \sqrt{x - 7}$

7) $4 = \sqrt{r - 3}$

8) $\sqrt{x - 5} = 7$

9) $12 = \sqrt{x - 4}$

10) $\sqrt{m + 5} = 8$

11) $7\sqrt{5a} = 35$

12) $6\sqrt{2x} = 48$

13) $2 = \sqrt{6x - 32}$

14) $\sqrt{304 - 4x} = 4$

15) $\sqrt{r + 2} - 8 = 6$

16) $-21 = -7\sqrt{r + 9}$

17) $60 = 6\sqrt{5v}$

18) $x = \sqrt{40 - 3x}$

19) $\sqrt{90 - 27a} = 3a$

20) $\sqrt{-8n + 88} = 4$

21) $\sqrt{15r - 5} = 4r - 3$

22) $\sqrt{-64 + 32x} = 4x$

23) $\sqrt{4x + 15} = \sqrt{2x + 11}$

24) $\sqrt{12v} = \sqrt{15v - 21}$

25) $\sqrt{9 - x} = \sqrt{x - 3}$

26) $\sqrt{6m + 34} = \sqrt{8m + 34}$

27) $\sqrt{7r + 32} = \sqrt{-8 - 3r}$

28) $\sqrt{4k + 10} = \sqrt{2 - 4k}$

29) $-20\sqrt{x - 13} = -40$

30) $\sqrt{90 - 2x} = \sqrt{\dfrac{x}{4}}$

Answers of Worksheets

Evaluating Function

1) $h(x) = -8x + 3$
2) $k(a) = 2a - 14$
3) $d(t) = 11t$
4) $f(x) = \frac{5}{12}x - \frac{7}{12}$
5) $m(n) = 24n - 210$
6) $c(p) = p^2 - 5p + 10$
7) -13
8) 14
9) -3
10) 22
11) 9
12) 26
13) 2
14) 7
15) -27
16) 8.2
17) 15
18) 15
19) 74
20) 75
21) $-1\frac{7}{8}$
22) $-3\frac{8}{9}$
23) 87
24) -28
25) 14
26) $-\frac{14b+9}{3b}$
27) $12a - 50$
28) $-2x + 18$
29) $2x^2 + 16$
30) $16x^4 - 20$

Adding and Subtracting Functions

1) -2
2) 1
3) -4
4) $-3x^2 - 6x - 19$
5) -43
6) 15
7) $-7\frac{2}{3}$
8) $4a^2 + 8a - 4$
9) $-18x^2 - 18x - 12$
10) $-2t^2 - 11t + 1$
11) $5x^4 - 5x^2 + 9$
12) $-81x^8 + 22$

Multiplying and Dividing Functions

1) -55
2) -70
3) 153
4) -1
5) $4\frac{4}{7}$
6) 2
7) 84
8) 2
9) 0
10) -62
11) $6x^4 - 17x^3 + 5x^2 + 3x - 1$
12) $5x^8 - 2x^6 - 10x^4 + 4x^2$

Composition of Functions

1) -13
2) -1
3) 26
4) -10
5) -17
6) 10
7) $\frac{25}{8}$
8) $\frac{1}{8}$
9) 8
10) $\frac{5}{8}$
11) $\frac{49}{8}$
12) $-\frac{1}{2}(x^2 - \frac{3}{2})$
13) 1
14) -2
15) 2
16) 6
17) -1
18) $-2x$

19) 3

20) -9

21) 2

22) $\sqrt{13}$

23) -7

24) -18

Quadratic Equations

1) $x^2 + 2x - 24$

2) $x^2 + 12x + 35$

3) $x^2 + 2x - 48$

4) $x^2 - 7x - 18$

5) $x^2 - 15x + 56$

6) $3x^2 - 7x - 6$

7) $4x^2 + 5x - 6$

8) $4x^2 - x - 5$

9) $7x^2 - 41x - 6$

10) $15x^2 - 12x - 3$

11) $(x-4)(x+2)$

12) $(x+5)(x+3)$

13) $(x-6)(x+4)$

14) $(x-3)(x-7)$

15) $(x+3)(x+7)$

16) $(4x+5)(x+1)$

17) $(5x-2)(x+3)$

18) $(5x-3)(x+4)$

19) $(2x+5)(x+1)$

20) $3(x-2)(3x-1)$

21) $x = -6, x = 3$

22) $x = -1, x = -8$

23) $x = -2, x = -5$

24) $x = 1, x = -2$

25) $x = 3, x = -4$

26) $x = -3, x = -8$

27) $x = -4, x = -\frac{1}{2}$

28) $x = 4, x = -7$

29) $x = 3, x = -4$

30) $x = -3, x = 2$

Solving quadratic equations

1) $\{-9, 1\}$

2) $\{-6, -7\}$

3) $\{8, -3\}$

4) $\{6, 4\}$

5) $\{-2, -12\}$

6) $\{-\frac{4}{5}, -7\}$

7) $\{-\frac{5}{4}, -\frac{1}{6}\}$

8) $\{-\frac{7}{2}, -8\}$

9) $\{-6, -5\}$

10) $\{-\frac{1}{6}, -8\}$

11) $\{8, 0\}$

12) $\{4, -4\}$

13) $\{2, 1\}$

14) $\{-4, -1\}$

15) $\{1, -9\}$

16) $\{2, -12\}$

17) $\{2, -8\}$

18) $\{-3, -6\}$

19) $\{-4, -9\}$

20) $\{5, -3\}$

21) $\{-5, -3\}$

22) $\{1, 3\}$

23) $\{\frac{6}{5}, \frac{3}{2}\}$

24) $\{\frac{6}{7}, 0\}$

25) $\{-\frac{7}{2}, 2\}$

26) $\{\frac{3}{5}, 2\}$

27) $\{-\frac{4}{3}, -4\}$

28) $\{-8, -7\}$

29) $\{4, -1\}$

30) $\{3, -6\}$

31) $\{3, 8\}$

32) $\{\frac{15}{17}, 0\}$

Quadratic formula and the discriminant

1) 576

2) 68

3) 8

4) -11

5) 28

6) 116

7) -11

8) 52

9) 40

10) 69

11) 64

12) 169

13) 336

14) 33

15) 316

16) 132

17) 104

18) 16

19) 1

20) 33

21) $0, \ one\ real\ solution$

22) $0, \ one\ real\ solution$

23) $-783, \ no\ solution$

24) 0, *one real solution* 26) 0, *one real solution* 28) 0, *one real solution*

25) −127, *no solution* 27) −703, *no solution*

Graphing quadratic functions

1) $(-3, 2), x = -3$

2) $(3, -2), x = 3$

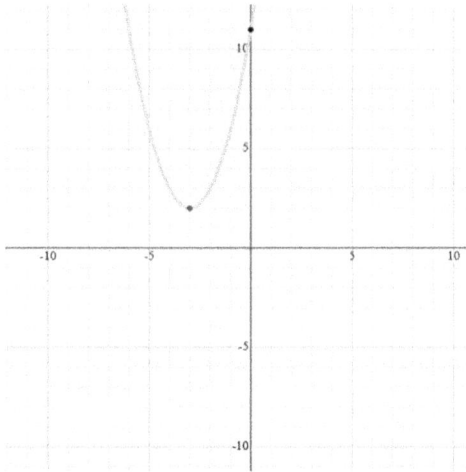

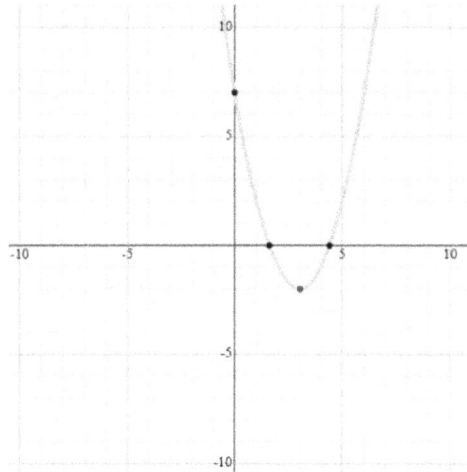

3) $(4, 6), x = 4$

4) $(-1, 12), x = -1$

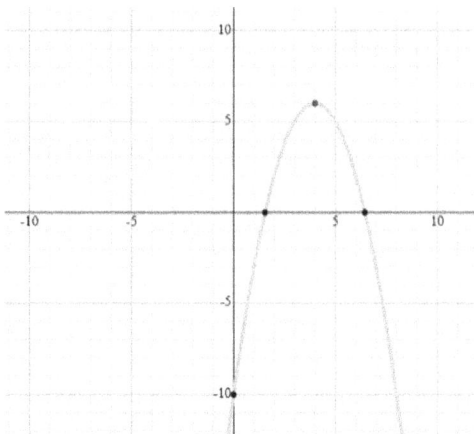

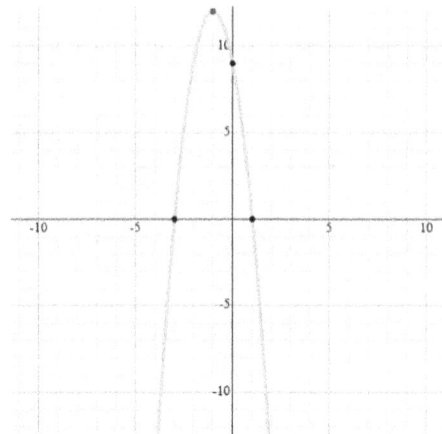

Quadratic inequalities

1) $-5 < x < 5$

2) $-4 < x < -2$

3) no solution

4) $x < -4 \ or \ x > -4$

5) $x \le -1 \ or \ x \ge 10$

6) $x < -5 \ or \ x > -5$

7) no solution

8) $-2 \le x \le 7$

9) $x \le -1 \ or \ x \ge 7$

10) $-9 < x < 1$

11) $x < -12 \ or \ x > 6$

12) $-3 < x < 4$

13) no solution

14) $x = 6$

15) $x \le 7 or \ x \ge 9$

16) $x \le 5 or \ x \ge 11$

17) $-9 \le x \le 9$

18) $x \le 3 \ or \ x \ge 14$

19) no solution

20) $x < -3$ or $x > 4$

21) no solution

22) $x \leq 1$ or $x \geq 5$

23) $x < -3$ or $x > 5$

24) $x \geq 3$

25) $x < -6$ or $x > 3$

26) $-6 < x < 1$

27) no solution

28) $x \leq -1$ or $x \geq 6$

29) $-5 < x < -3$

30) no solution

31) $x \leq -4$ or $x \geq 3$

32) $x \leq -1$ or $x \geq 3$

Domain and range of radical functions

1) domain: $x \geq -8$

   range: $y \geq -7$

2) domain: {all real numbers}

   range: {all real numbers}

3) domain: $x \geq 3$

   range: $y \geq 3$

4) domain: {all real numbers}

   range: {all real numbers}

5) domain: $x \geq -5$

   range: $y \geq 6$

6) domain: {all real numbers}

   range: {all real numbers}

7) domain: {all real numbers}

   range: $y \geq 8\sqrt{2} + 3$

8) domain: {all real numbers}

   range: {all real numbers}

9) domain: $x \geq -2$

   range: $y \geq -3$

10) domain: {all real numbers}

   range: {all real numbers}

11) domain: $x \leq -2$

   range: $y \geq 5$

12) domain: {all real numbers}

   range: {all real numbers}

13) domain: $x \geq 5$

   range: $y \geq -2$

14) domain: {all real numbers}

   range: {all real numbers}

15)

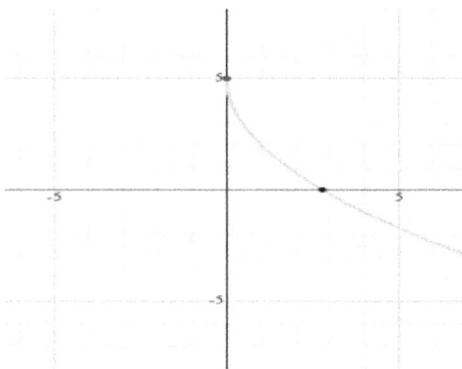

16)

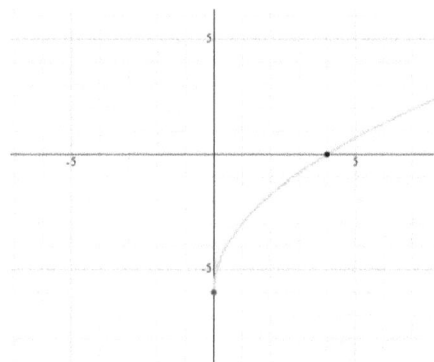

Solving radical equations

1) {81}

2) {36}

3) {16}

4) $\{\frac{1}{4}\}$

5) {321}

6) {43}

7) {19}

8) {54}

9) {148}

10) {59}

11) {5}

12) {32}

13) {6}

14) {72}

15) {194}

16) {0}

17) {20}

18) {5}

19) {2}

20) {9}

21) {2}

22) no solution

23) {−2}

24) {7}

25) {6}

26) {0}

27) {−4}

28) {−1}

29) {17}

30) {40}

Chapter 10 :

Geometry and Solid Figures

Topics that you will practice in this chapter:

- ✓ Angles
- ✓ Pythagorean Relationship
- ✓ Triangles
- ✓ Polygons
- ✓ Trapezoids
- ✓ Circles
- ✓ Cubes
- ✓ Rectangular Prism
- ✓ Cylinder
- ✓ Pyramids and Cone

Mathematics is, as it were, a sensuous logic, and relates to philosophy as do the arts, music, and plastic art to poetry. — *K. Shegel*

Angles

✎ What is the value of x in the following figures?

1)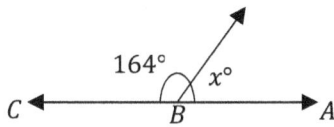

2)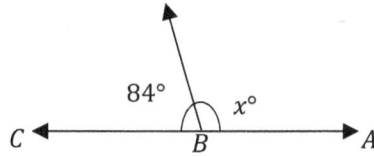

3)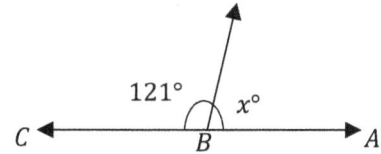

4)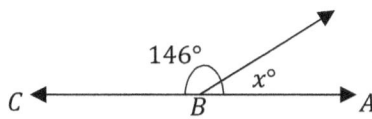

5)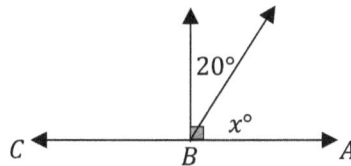

6)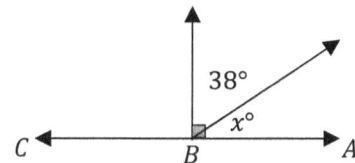

✎ Calculate.

7) Two supplement angles have equal measures. What is the measure of each angle? _____

8) The measure of an angle is seven fifth the measure of its supplement. What is the measure of the angle? _____

9) Two angles are complementary and the measure of one angle is 24 less than the other. What is the measure of the smaller angle? _____

10) Two angles are complementary. The measure of one angle is one fifth the measure of the other. What is the measure of the bigger angle? _____

11) Two supplementary angles are given. The measure of one angle is 40° less than the measure of the other. What does the smaller angle measure? _____

Pythagorean Relationship

✍ **Do the following lengths form a right triangle?**

1) 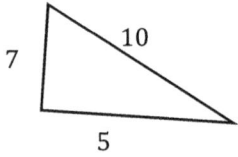	2) 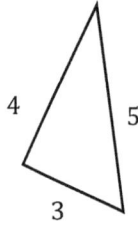	3) 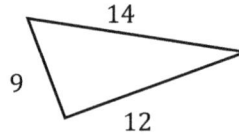	4) 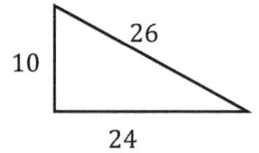
5) 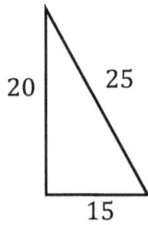	6) 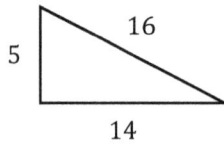	7) 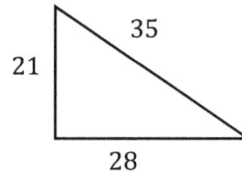	8) 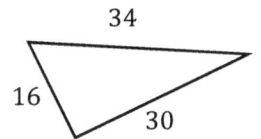

✍ **Find the missing side?**

9) 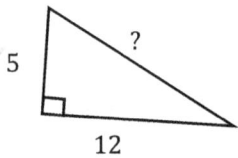	10) 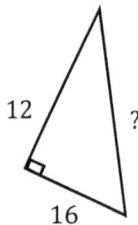	11) 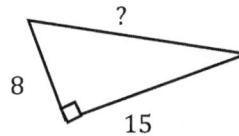	12) 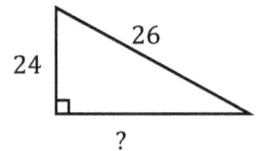
13) 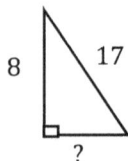	14) 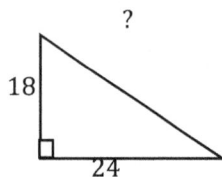	15) 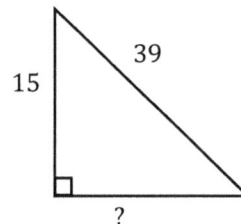	16) 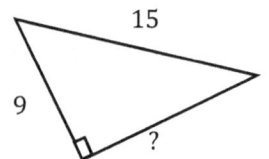

Triangles

✎ **Find the measure of the unknown angle in each triangle.**

1)
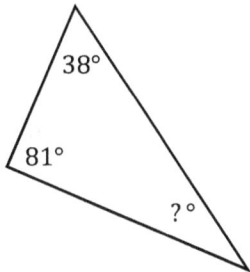

2)
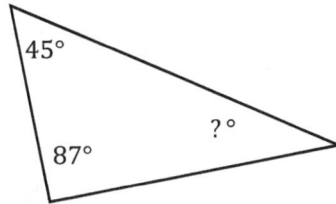

3)
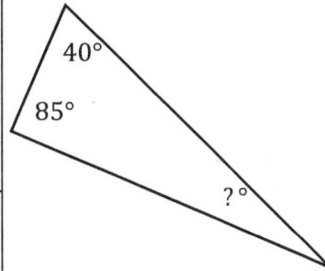

4)
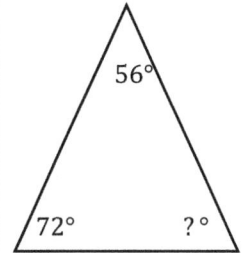

5)
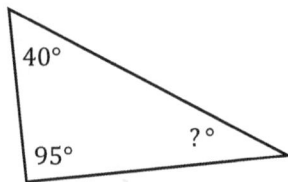

6)
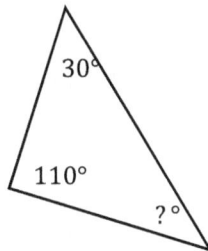

7)
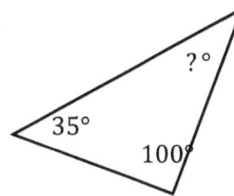

8)
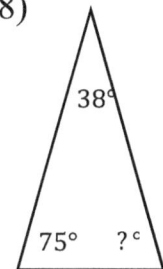

✎ **Find area of each triangle.**

9)
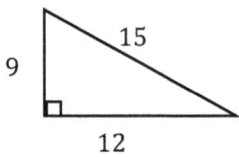

10)
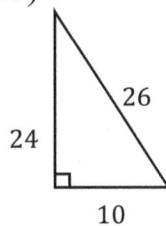

11)
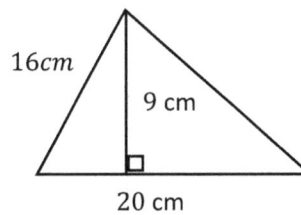

12)
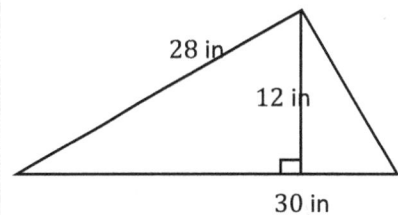

Polygons

✍ Find the perimeter of each shape.

1)

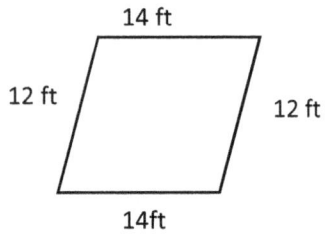

14 ft
12 ft
12 ft
14ft

2)

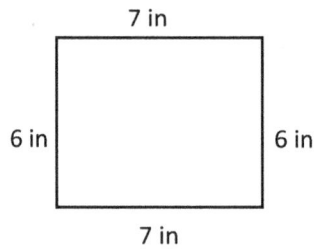

7 in
6 in
6 in
7 in

3)

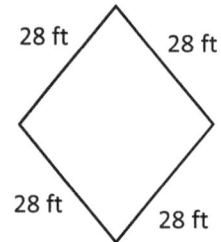

28 ft
28 ft
28 ft
28 ft

4) Square

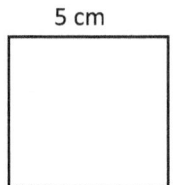

5 cm

5) Regular hexagon

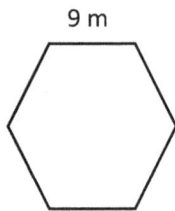

9 m

6)

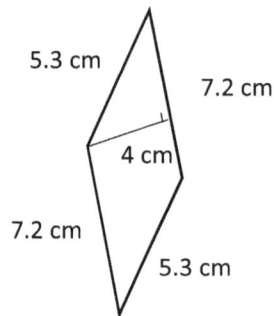

5.3 cm
7.2 cm
4 cm
7.2 cm
5.3 cm

7) Parallelogram

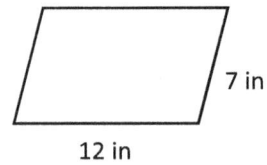

7 in
12 in

8) Square

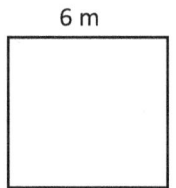

6 m

✍ Find the area of each shape.

9) Parallelogram

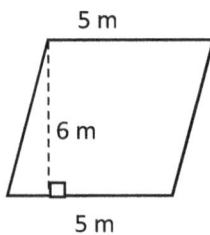

5 m
6 m
5 m

10) Rectangle

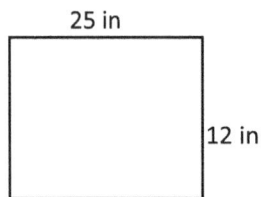

25 in
12 in

11) Rectangle

16 km
10 km

12) Square

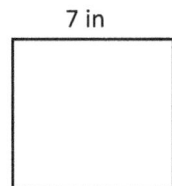

7 in

Trapezoids

✎ **Find the area of each trapezoid.**

1)
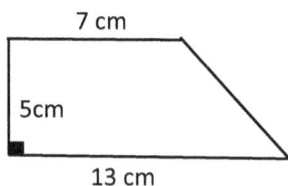

2)
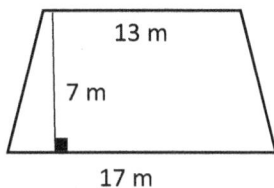

3)
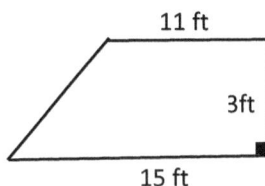

4)
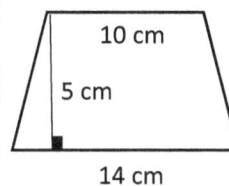

5)
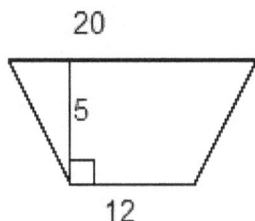

6)
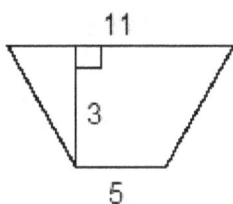

7)
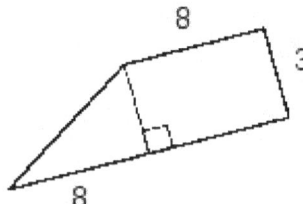

8)
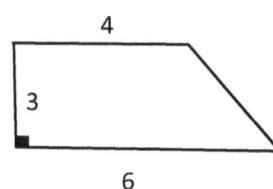

✎ **Calculate.**

1) A trapezoid has an area of 45 cm² and its height is 5 cm and one base is 5 cm. What is the other base length? _____

2) If a trapezoid has an area of 99 ft² and the lengths of the bases are 8 ft and 10 ft, find the height? _____

3) If a trapezoid has an area of 126 m² and its height is 14 m and one base is 6 m, find the other base length? _____

4) The area of a trapezoid is 440 ft² and its height is 22 ft. If one base of the trapezoid is 15 ft, what is the other base length?

_____

Circles

✍ **Find the area of each circle.** ($\pi = 3.14$)

1)	2)	3)	4)	5)	6)
2.5 in	5cm	9 ft	2 m	11 cm	10 miles

7)	8)	9)	10)	11)	12)
13 in	3 ft	4 m	12 cm	7 miles	8 ft

✍ **Complete the table below.** ($\pi = 3.14$)

Circle No.	Radius	Diameter	Circumference	Area
1	1 in	2 in	6.28 in	3.14 in^2
2		10 m		
3				28.26 ft^2
4			47.1 mi	
5		11 km		
6	7 cm			
7		12 ft		
8				314 m^2
9			56.52 in	
10	4.5 ft			

Cubes

✏ **Find the volume of each cube.**

1)	2)	3)	4)	5)	6)
	2 cm	6 ft	11 m	13 in	7 miles

7)	8	9)	10)	11)	12)
1.2 km	9 cm	2.1 ft	12 mm	0.2 in	0.1 km

✏ **Find the surface area of each cube.**

13)	14)	15)	16)	17)	18)
	7 m	5 ft	4.5 mm	1.1 km	11 cm

Rectangular Prism

✏ **Find the volume of each Rectangular Prism.**

1)

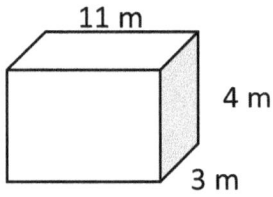

2)

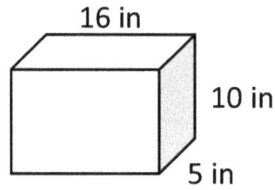

3)

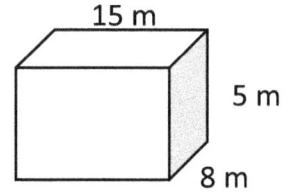

4)

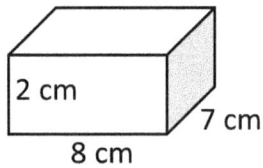

5)

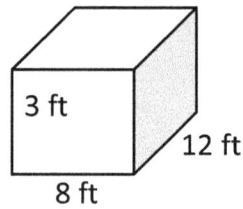

6)

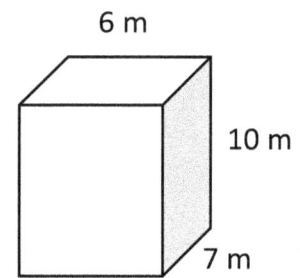

✎ **Find the surface area of each Rectangular Prism.**

7)

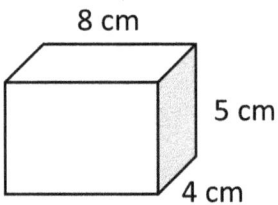

8)

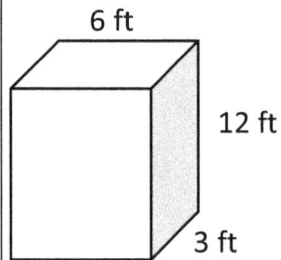

9)

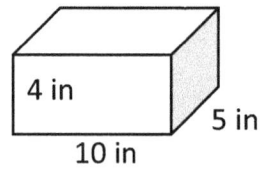

10)

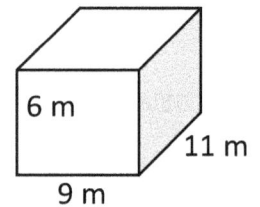

Cylinder

✎ **Find the volume of each Cylinder. Round your answer to the nearest tenth.** ($\pi = 3.14$)

1)

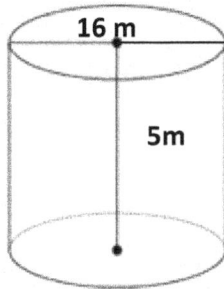

16 m

5m

2)

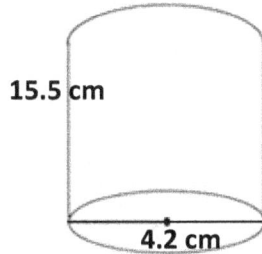

15.5 cm

4.2 cm

3)

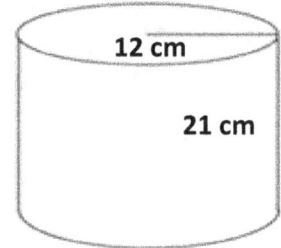

12 cm

21 cm

4)

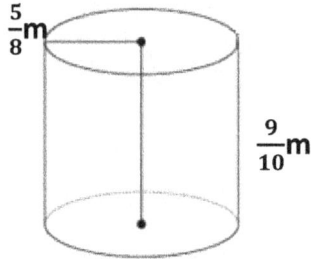

$\frac{5}{8}$m

$\frac{9}{10}$m

5)

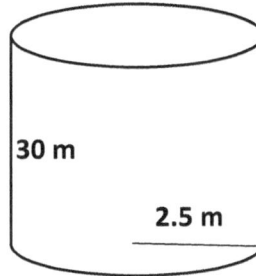

30 m

2.5 m

6)

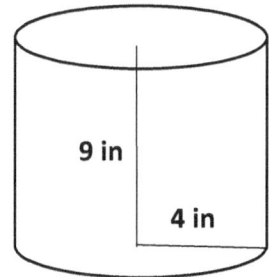

9 in

4 in

✎ **Find the surface area of each Cylinder.** ($\pi = 3.14$)

7)

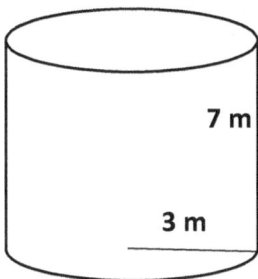

7 m

3 m

8)

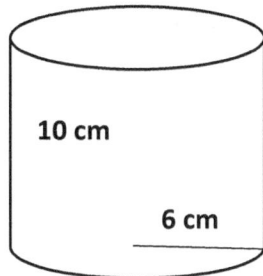

10 cm

6 cm

9)

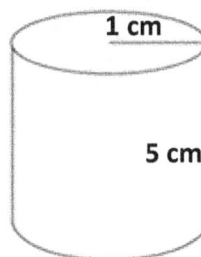

1 cm

5 cm

10)

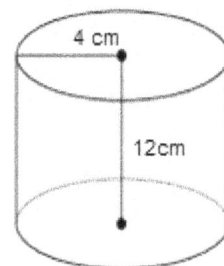

4 cm

12cm

Pyramids and Cone

🖎 **Find the volume of each Pyramid and Cone.** ($\pi = 3.14$)

1)
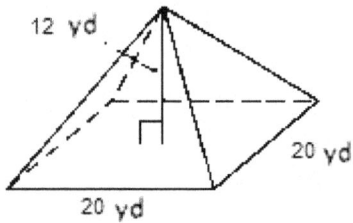

2)
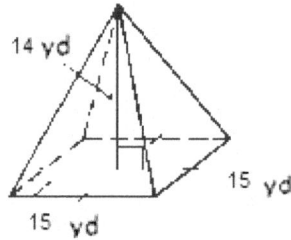

3)
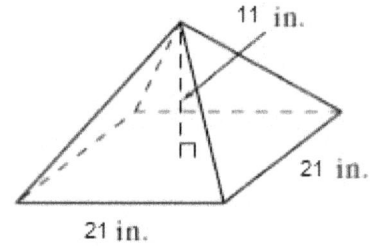

4)
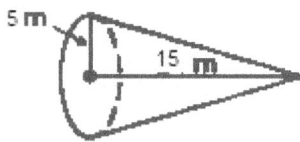

5)
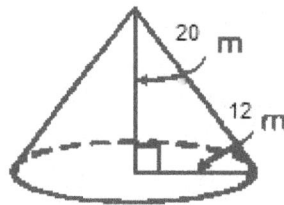

6)
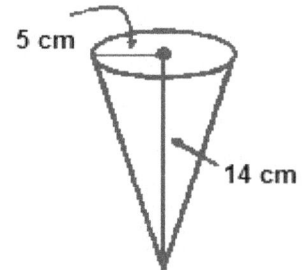

🖎 **Find the surface area of each Pyramid and Cone.** ($\pi = 3.14$)

7)

8)

9)

10)

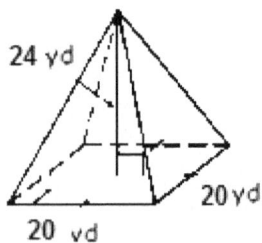

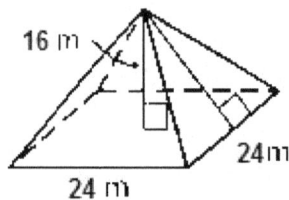

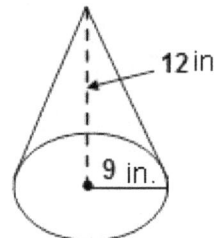

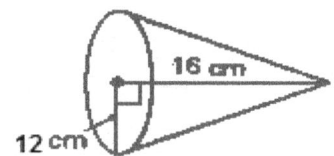

Answers of Worksheets

Angles

1) 16°   4) 34°   7) 90°   10) 75°

2) 96°   5) 70°   8) 75°   11) 70°

3) 59°   6) 52°   9) 33°

Pythagorean Relationship

1) No   5) Yes   9) 13   13) 15

2) Yes   6) No   10) 20   14) 30

3) No   7) Yes   11) 17   15) 36

4) Yes   8) Yes   12) 10   16) 12

Triangles

1) 60°   5) 45°   9) 54 $square\ unites$

2) 48°   6) 40°   10) 120 $square\ unites$

3) 55°   7) 45°   11) 90 $square\ unites$

4) 52°   8) 67°   12) 180 $square\ unites$

Polygons

1) 52 ft 5) 54 m 9) 30 m^2

2) 26 in 6) 25 cm 10) 300 in^2

3) 112 ft 7) 38 in 11) 160 km^2

4) 20 cm 8) 24 m 12) 49 in^2

Trapezoids

1) 50 cm^2 4) 60 cm^2 7) 36

2) 105 m^2 5) 80   8) 15

3) 39 ft^2 6) 24

Calculate

1) 13 cm 2) 11 ft 3) 12 m 4) 25 ft

Circles

1) 19.63 in^2 5) 379.94 cm^2 9) 12.56 m^2

2) 78.5 cm^2 6) 314 $miles^2$ 10) 113.04 cm^2

3) 254.34 ft^2 7) 132.67 in^2 11) 38.47 $miles^2$

4) 12.56 m^2 8) 7.07 ft^2 12) 50.24 ft^2

Circle No.	Radius	Diameter	Circumference	Area
1	1 in	2 in	6.28 in	3.14 in^2
2	5 m	10 m	31.4 m	78.5 m^2
3	3 ft	6 ft	18.84 ft	28.26 ft^2
4	7.5 miles	15 mi	47.1 mi	176.63 mi^2
5	5.5 km	11 km	34.54 km	94.99 km^2
6	7 cm	14 cm	43.96 cm	153.86 cm^2
7	6 ft	12 ft	37.68 feet	113.04 ft^2
8	10 m	20 m	62.8 m	314 m^2
9	9 in	18 in	56.52 in	254.34 in^2
10	4.5 ft	9 ft	28.26 ft	63.585 ft^2

Cubes

1) 12
2) 8 cm^3
3) 216 ft^3
4) 1,331 m^3
5) 2,197 in^3

6) 343 $miles^3$
7) 1.728 km^3
8) 729 cm^3
9) 9.261 ft^3
10) 1,728 mm^3

11) 0.008 in^3
12) 0.001 km^3
13) 27
14) 294 m^2
15) 150 ft^2

16) 121.5 mm^2
17) 7.26 km^2
18) 726 cm^2

Rectangular Prism

1) 132 m^3
2) 800 in^3
3) 600 m^3

4) 112 cm^3
5) 288 ft^3
6) 420 m^3

7) 184 cm^2
8) 252 ft^2
9) 220 in^2

10) 438 m^2

Cylinder

1) 1,004.8 m^3
2) 214.6 cm^3
3) 9,495.4 cm^3

4) 1.1 m^3
5) 588.8 m^3
6) 452.2 in^3

7) 188.4 m^2
8) 602.9 cm^2
9) 37.7 cm^2

10) 401.9 m^2

Pyramids and Cone

1) 1,600 yd^3
2) 1,050 yd^3
3) 1,617 in^3

4) 392.5 m^3
5) 3,014.4 m^3
6) 366.33 cm^3

7) 1,440 yd^2
8) 1,536 m^2
9) 678.24 in^2

10) 1,205.76 cm^2

Chapter 11 :

Statistics and Probability

Topics that you will practice in this chapter:

- ✓ Mean and Median
- ✓ Mode and Range
- ✓ Histograms
- ✓ Stem–and–Leaf Plot
- ✓ Pie Graph
- ✓ Probability Problems
- ✓ Factorials
- ✓ Combinations and Permutation

Mathematics is no more computation than typing is literature.
– John Allen Paulos

Mean and Median

🖎 **Find Mean and Median of the Given Data.**

1) 8, 7, 14, 4, 8

2) 14, 8, 25, 19, 16, 33, 11

3) 23, 18, 15, 12, 17

4) 34, 14, 10, 15, 6, 11

5) 10, 19, 6, 8, 32, 20, 17

6) 17, 26, 39, 69, 20, 6

7) 40, 38, 18, 11, 9, 2, 7, 32, 41

8) 24, 21, 31, 12, 33, 32, 22

9) 16, 14, 20, 41, 15, 20, 38, 4

10) 20, 20, 30, 18, 6, 28, 12, 46

11) 12, 7, 10, 11, 16, 22

12) 10, 29, 27, 12, 2, 15, 10, 3

🖎 **Calculate.**

13) In a javelin throw competition, five athletics score 56, 34, 62, 23 and 19 meters. What are their Mean and Median? _____

14) Eva went to shop and bought 8 apples, 14 peaches, 6 bananas, 4 pineapples and 12 melons. What are the Mean and Median of her purchase? _____

15) Bob has 17 black pen, 19 red pen, 14 green pens, 20 blue pens and 5 boxes of yellow pens. If the Mean and Median are 19 respectively, what is the number of yellow pens in each box? _____

Mode and Range

✍ **Find Mode and Rage of the Given Data.**

1) 4, 3, 7, 3, 3, 4

Mode: _____   Range: _____

2) 18, 18, 24, 26, 18, 8, 14, 22

Mode: _____   Range: _____

3) 8, 8, 8, 16, 19, 22, 20, 9, 13

Mode: _____   Range: _____

4) 24, 24, 14, 28, 20, 18, 20, 24

Mode: _____   Range: _____

5) 6, 21, 27, 24, 27, 27

Mode: _____   Range: _____

6) 21, 8, 8, 7, 8, 12, 10, 22, 18, 13

Mode: _____   Range: _____

7) 7, 4, 4, 6, 13, 13, 13, 0, 2, 2

Mode: _____   Range: _____

8) 5, 8, 5, 14, 12, 14, 3, 5, 18

Mode: _____   Range: _____

9) 7, 7, 7, 12, 7, 3, 8, 16, 3, 17

Mode: _____   Range: _____

10) 15, 15, 19, 16, 4, 16, 10, 15

Mode: _____   Range: _____

11) 6, 6, 5, 6, 42, 13, 19, 2

Mode: _____   Range: _____

12) 8, 8, 9, 8, 9, 4, 34, 22

Mode: _____   Range: _____

✍ **Calculate.**

13) A stationery sold 12 pencils, 56 red pens, 24 blue pens, 20 notebooks, 12 erasers, 21 rulers and 11 color pencils. What are the Mode and Range for the stationery sells?

Mode: _____   Range: _____

14) In an English test, eight students score 10, 15, 15, 18 18, 16, 15 and 15. What are their Mode and Range? _____

15) What is the range of the first 6 even numbers greater than 8?

_____

Times Series

🖎 **Use the following Graph to complete the table.**

Day	Distance (km)
1	
2	

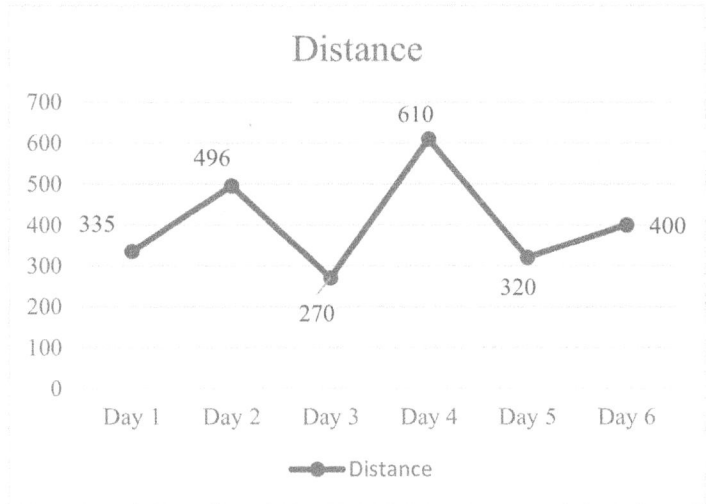

Distance

700
610
600
496
500
400 335
300
200 270 320 400
100
0

Day 1   Day 2   Day 3   Day 4   Day 5   Day 6

━●━ Distance

The following table shows the number of births in the US from 2007 to 2012 (in millions).

Year	Number of births (in millions)
2007	4.15
2008	3.70
2009	3.45
2010	3.20
2011	1.75
2012	2.98

Draw a Time Series for the table.

Stem–and–Leaf Plot

✎ **Make stem ad leaf plots for the given data.**

1) $24, 26, 29, 20, 53, 27, 51, 55, 36, 21, 37, 30$

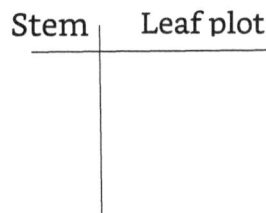

Stem | Leaf plot

2) $11, 59, 66, 14, 18, 19, 59, 65, 69, 61, 68, 65$

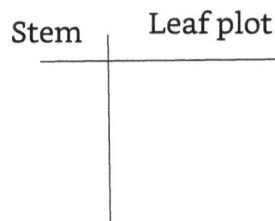

Stem | Leaf plot

3) $121, 55, 66, 54, 112, 128, 63, 125, 59, 123, 68, 119$

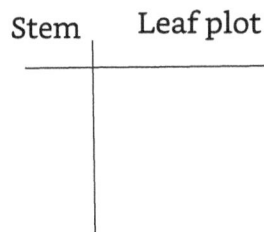

Stem | Leaf plot

4) $51, 32, 100, 56, 84, 36, 107, 56, 85, 39, 56, 106, 89$

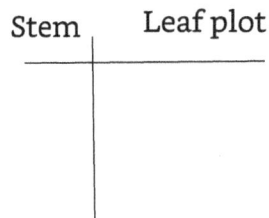

Stem | Leaf plot

5) $33, 89, 19, 87, 81, 16, 11, 30, 86, 35, 17, 35, 13$

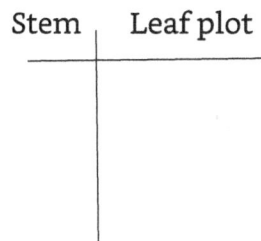

Stem | Leaf plot

6) $60, 92, 22, 25, 67, 93, 95, 62, 21, 64, 98, 29$

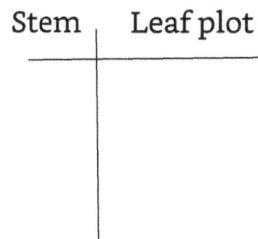

Stem | Leaf plot

Pie Graph

The circle graph below shows all Robert's expenses for last month. Robert spent $140 on his hobbies last month.

Answer following questions based on the Pie graph.

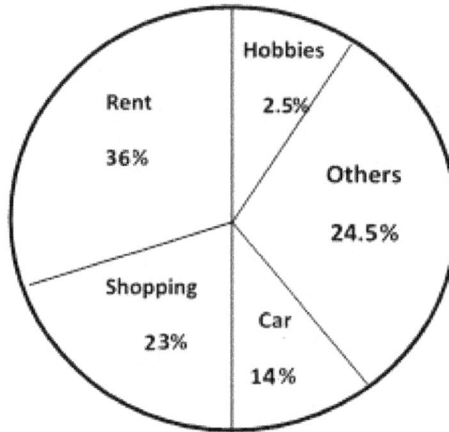

1) How much was Robert's total expenses last month? _____

2) How much did Robert spend on his car last month? _____

3) How much did Robert spend for shopping last month? _____

4) How much did Robert spend on his rent last month? _____

5) What fraction is Robert's expenses for his rent and car out of his total

   expenses last month? _____

Probability Problems

✎ **Calculate.**

1) A number is chosen at random from 1 to 10. Find the probability of selecting number 6 or smaller numbers. _____

2) Bag A contains 18 red marbles and 6 green marbles. Bag B contains 16 black marbles and 8 orange marbles. What is the probability of selecting a green marble at random from bag A? What is the probability of selecting a black marble at random from Bag B? _____

3) A number is chosen at random from 1 to 20. What is the probability of selecting multiples of 4? _____

4) A card is chosen from a well-shuffled deck of 52 cards. What is the probability that the card will be a queen? _____

5) A number is chosen at random from 1 to 15. What is the probability of selecting a multiple of 3 or 5? _____

A spinner numbered 1–8, is spun once. What is the probability of spinning …?

6) an Odd number? _____   7) a multiple of 2? _____

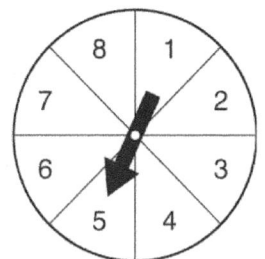

8) a multiple of 5? _____   9) number 10? _____

Factorials

✍ **Determine the value for each expression.**

1) $4! + 0! =$

2) $2! + 5! =$

3) $(2!)^2 =$

4) $5! - 3! =$

5) $6! - 3! + 10 =$

6) $3! \times 4 - 15 =$

7) $(2! + 3!)^2 =$

8) $(4! - 3!)^2 =$

9) $(3! \, 0!)^2 - 10 =$

10) $\dfrac{10!}{8!} =$

11) $\dfrac{6!}{4!} =$

12) $\dfrac{6!}{5!} =$

13) $\dfrac{15!}{13!} =$

14) $\dfrac{n!}{(n-3)!} =$

15) $\dfrac{(n+2)!}{n!} =$

16) $\dfrac{(2+2!)^3}{2!} =$

17) $\dfrac{5(n+2)!}{(n+1)!} =$

18) $\dfrac{22!}{20!4!} =$

19) $\dfrac{13!}{11!3!} =$

20) $\dfrac{9 \times 210!}{3(7 \times 30)!} =$

21) $\dfrac{32!}{31!2!} =$

22) $\dfrac{11!12!}{10!13!} =$

23) $\dfrac{16!15!}{14!14!} =$

24) $\dfrac{(5 \times 3)!}{0!14!} =$

25) $\dfrac{4!(5n-2)!}{(5n)!} =$

26) $\dfrac{4n(4n+7)!}{(4n+8)!} =$

27) $\dfrac{(n-2)!(n+1)}{(n+2)!} =$

Combinations and Permutations

✎ **Calculate the value of each.**

1) 6! = _____

2) 2! × 5! = _____

3) 3 × 4! = _____

4) 5! + 3! = _____

5) 7! = _____

6) 4! = _____

7) 3! + 3! = _____

8) 7! − 5! = _____

✎ **Find the answer for each word problems.**

9) Susan is baking cookies. She uses sugar, butter, Vanilla, eggs and flour. How many different orders of ingredients can she try? _____

10) Albert is planning for his vacation. He wants to go to museum, watch a movie, go to the beach, play the game and play football. How many ways of ordering are there for him? _____

11) How many 4-digit numbers can be named using the digits 3, 4, 5, and 6 without repetition? _____

12) In how many ways can 5 boys be arranged in a straight line? _____

13) In how many ways can 6 athletes be arranged in a straight line? _____

14) A professor is going to arrange her 7 students in a straight line. In how many ways can she do this? _____

15) How many code symbols can be formed with the letters for the word GAMES? _____

16) In how many ways a team of 7 basketball players can choose a captain and co-captain? _____

Answers of Worksheets

Mean and Median

1) Mean: 8.2, Median: 8
2) Mean: 18, Median: 16
3) Mean: 17, Median: 17
4) Mean: 15, Median: 12.5
5) Mean: 16, Median: 17

6) Mean: 29.5, Median: 23
7) Mean: 22, Median: 18
8) Mean: 25, Median: 24
9) Mean: 21, Median: 18
10) Mean: 22.5, Median: 20

11) Mean: 13, Median: 11.5
12) Mean: 13.5, Median: 11
13) Mean: 38.8, Median: 34
14) Mean: 8.8, Median: 8
15) 5

Mode and Range

1) Mode: 3, Range: 4
2) Mode: 18, Range: 18
3) Mode: 8, Range: 14
4) Mode: 24, Range: 14
5) Mode: 27, Range: 21

6) Mode: 8, Range: 15
7) Mode: 13, Range: 13
8) Mode: 5, Range: 15
9) Mode: 7, Range: 14
10) Mode: 15, Range: 15

11) Mode: 6, Range: 40
12) Mode: 8, Range: 30
13) Mode: 12, Range: 45
14) Mode: 15, Range: 8
15) 10

Time series

Day	Distance (km)
1	335
2	496
3	270
4	610
5	320
6	400

Number of Births

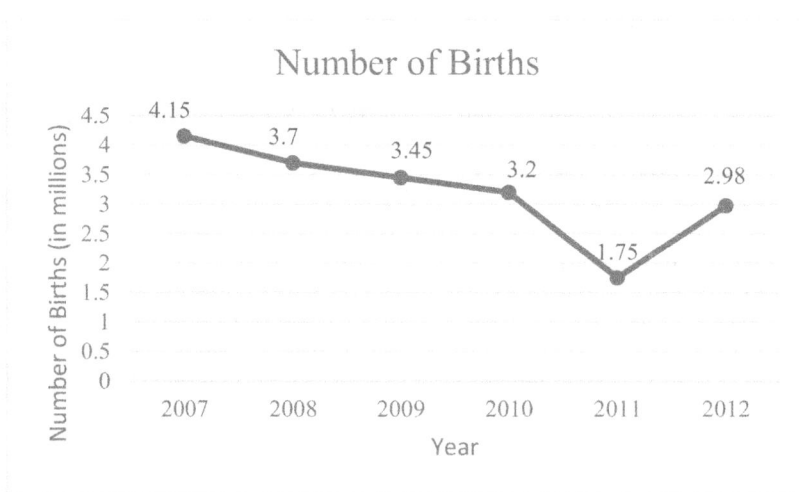

Stem–And–Leaf Plot

1)

Stem	leaf
2	0 1 4 6 7 9
3	0 6 7
5	1 3 5

2)

Stem	leaf
1	1 4 8 9
5	9 9
6	1 5 5 6 8 9

3)

Stem	leaf
5	4 5 9
6	3 6 8
11	2 9
12	1 3 5 8

4)

Stem	leaf
3	2 6 9
5	1 6 6 6
8	4 5 9
10	0 6 7

5)

Stem	leaf
1	1 3 6 7 9
3	0 3 5 5
8	1 6 7 9

6)

Stem	leaf
2	2 1 5 9
6	0 2 4 7
9	2 3 5 8

Pie Graph

1) $5,600
2) $784
3) $1,288
4) $2,016
5) $\frac{1}{2}$

Probability Problems

1) $\frac{3}{5}$
2) $\frac{1}{4}, \frac{2}{3}$
3) $\frac{1}{4}$
4) $\frac{1}{13}$
5) $\frac{7}{15}$
6) $\frac{1}{2}$
7) $\frac{1}{2}$
8) $\frac{1}{8}$
9) 0

Factorials

1) 25
2) 122
3) 4
4) 114
5) 724
6) 9
7) 64
8) 324
9) 26
10) 90
11) 30
12) 6
13) 210
14) $n(n-1)(n-2)$
15) $(n+1)(n+2)$
16) 32
17) $5(n+2)$
18) 19.25
19) 26
20) 3
21) 16
22) $\frac{11}{13}$
23) 3,600
24) 15
25) $\frac{24}{5n(5n-1)}$
26) $\frac{n}{(n+2)}$
27) $\frac{1}{n(n-1)(n+2)}$

Combinations and Permutations

1) 720
2) 240
3) 72
4) 126
5) 5,040
6) 24
7) 12
8) 4,920
9) 120
10) 120
11) 24
12) 120
13) 720
14) 5,040
15) 120
16) 42

Chapter 12 : ISEE Upper-Level Practice Tests

The Independent School Entrance Exam (ISEE) is a standardized test developed by the Educational Records Bureau for its member schools as part of their admission process. There are currently four Levels of the ISEE:

- ✓ Primary Level (entering Grades 2 - 4)
- ✓ Lower Level (entering Grades 5 and 6)
- ✓ Middle Level (entering Grades 7 and 8)
- ✓ Upper-Level (entering Grades 9 - 12)

There are five sections on the ISEE Upper-Level Test:

- o Verbal Reasoning
- o Quantitative Reasoning
- o Reading Comprehension
- o Mathematics Achievement
- o and a 30-minute essay

ISEE Upper-Level tests use a multiple-choice format and contain two Mathematics sections:

Quantitative Reasoning

There are 37 questions in the Quantitative Reasoning section and students have 35 minutes to answer the questions. This section contains word problems and quantitative comparisons. The word problems require either no calculation or simple calculation. The quantitative comparison items present two quantities, (A) and (B), and the student needs to select one of the following four answer choices:

(A) The quantity in Column A is greater.

(B) The quantity in Column B is greater.

(C) The two quantities are equal.

(D) The relationship cannot be determined from the information given.

Mathematics Achievement

There are 47 questions in the Mathematics Achievement section and students have 40 minutes to answer the questions. Mathematics Achievement measures students' knowledge of Mathematics requiring one or more steps in calculating the answer.

In this book, we have reviewed Quantitative Reasoning and Mathematic Achievement topics being tested on the ISEE Upper-Level. In this section, there are two complete ISEE Upper-Level Quantitative Reasoning and Mathematics Achievement Tests. Let your student take these tests to see what score they will be able to receive on a real ISEE Upper-Level test.

Good Luck!

Time to Test

Time to refine your skill with a practice examination

Take a practice ISEE Upper-Level Math Test to simulate the test day experience. After you've finished, score your test using the answer key.

Before You Start

- You'll need a pencil and scratch papers to take the test.
- For each question, there are four possible answers. Choose which one is best.
- It's okay to guess. You won't lose any points if you're wrong.
- Use the answer sheet provided to record your answers.
- After you've finished the test, review the answer key to see where you went wrong.
- **Calculators are NOT allowed for the ISEE Upper-Level Test.**

Good Luck!

ISEE Upper-Level Practice Test Answer Sheets

Remove (or photocopy) these answer sheets and use them to complete the practice tests.

ISEE Upper-Level Practice Test

Quantitative Reasoning

1 Ⓐ Ⓑ Ⓒ Ⓓ	21 Ⓐ Ⓑ Ⓒ Ⓓ
2 Ⓐ Ⓑ Ⓒ Ⓓ	22 Ⓐ Ⓑ Ⓒ Ⓓ
3 Ⓐ Ⓑ Ⓒ Ⓓ	23 Ⓐ Ⓑ Ⓒ Ⓓ
4 Ⓐ Ⓑ Ⓒ Ⓓ	24 Ⓐ Ⓑ Ⓒ Ⓓ
5 Ⓐ Ⓑ Ⓒ Ⓓ	25 Ⓐ Ⓑ Ⓒ Ⓓ
6 Ⓐ Ⓑ Ⓒ Ⓓ	26 Ⓐ Ⓑ Ⓒ Ⓓ
7 Ⓐ Ⓑ Ⓒ Ⓓ	27 Ⓐ Ⓑ Ⓒ Ⓓ
8 Ⓐ Ⓑ Ⓒ Ⓓ	28 Ⓐ Ⓑ Ⓒ Ⓓ
9 Ⓐ Ⓑ Ⓒ Ⓓ	29 Ⓐ Ⓑ Ⓒ Ⓓ
10 Ⓐ Ⓑ Ⓒ Ⓓ	30 Ⓐ Ⓑ Ⓒ Ⓓ
11 Ⓐ Ⓑ Ⓒ Ⓓ	31 Ⓐ Ⓑ Ⓒ Ⓓ
12 Ⓐ Ⓑ Ⓒ Ⓓ	32 Ⓐ Ⓑ Ⓒ Ⓓ
13 Ⓐ Ⓑ Ⓒ Ⓓ	33 Ⓐ Ⓑ Ⓒ Ⓓ
14 Ⓐ Ⓑ Ⓒ Ⓓ	34 Ⓐ Ⓑ Ⓒ Ⓓ
15 Ⓐ Ⓑ Ⓒ Ⓓ	35 Ⓐ Ⓑ Ⓒ Ⓓ
16 Ⓐ Ⓑ Ⓒ Ⓓ	36 Ⓐ Ⓑ Ⓒ Ⓓ
17 Ⓐ Ⓑ Ⓒ Ⓓ	37 Ⓐ Ⓑ Ⓒ Ⓓ
18 Ⓐ Ⓑ Ⓒ Ⓓ	38 Ⓐ Ⓑ Ⓒ Ⓓ
19 Ⓐ Ⓑ Ⓒ Ⓓ	39 Ⓐ Ⓑ Ⓒ Ⓓ
20 Ⓐ Ⓑ Ⓒ Ⓓ	40 Ⓐ Ⓑ Ⓒ Ⓓ

Mathematics Achievement

1 Ⓐ Ⓑ Ⓒ Ⓓ	21 Ⓐ Ⓑ Ⓒ Ⓓ	41 Ⓐ Ⓑ Ⓒ Ⓓ
2 Ⓐ Ⓑ Ⓒ Ⓓ	22 Ⓐ Ⓑ Ⓒ Ⓓ	42 Ⓐ Ⓑ Ⓒ Ⓓ
3 Ⓐ Ⓑ Ⓒ Ⓓ	23 Ⓐ Ⓑ Ⓒ Ⓓ	43 Ⓐ Ⓑ Ⓒ Ⓓ
4 Ⓐ Ⓑ Ⓒ Ⓓ	24 Ⓐ Ⓑ Ⓒ Ⓓ	44 Ⓐ Ⓑ Ⓒ Ⓓ
5 Ⓐ Ⓑ Ⓒ Ⓓ	25 Ⓐ Ⓑ Ⓒ Ⓓ	45 Ⓐ Ⓑ Ⓒ Ⓓ
6 Ⓐ Ⓑ Ⓒ Ⓓ	26 Ⓐ Ⓑ Ⓒ Ⓓ	46 Ⓐ Ⓑ Ⓒ Ⓓ
7 Ⓐ Ⓑ Ⓒ Ⓓ	27 Ⓐ Ⓑ Ⓒ Ⓓ	47 Ⓐ Ⓑ Ⓒ Ⓓ
8 Ⓐ Ⓑ Ⓒ Ⓓ	28 Ⓐ Ⓑ Ⓒ Ⓓ	48 Ⓐ Ⓑ Ⓒ Ⓓ
9 Ⓐ Ⓑ Ⓒ Ⓓ	29 Ⓐ Ⓑ Ⓒ Ⓓ	49 Ⓐ Ⓑ Ⓒ Ⓓ
10 Ⓐ Ⓑ Ⓒ Ⓓ	30 Ⓐ Ⓑ Ⓒ Ⓓ	50 Ⓐ Ⓑ Ⓒ Ⓓ
11 Ⓐ Ⓑ Ⓒ Ⓓ	31 Ⓐ Ⓑ Ⓒ Ⓓ	
12 Ⓐ Ⓑ Ⓒ Ⓓ	32 Ⓐ Ⓑ Ⓒ Ⓓ	
13 Ⓐ Ⓑ Ⓒ Ⓓ	33 Ⓐ Ⓑ Ⓒ Ⓓ	
14 Ⓐ Ⓑ Ⓒ Ⓓ	34 Ⓐ Ⓑ Ⓒ Ⓓ	
15 Ⓐ Ⓑ Ⓒ Ⓓ	35 Ⓐ Ⓑ Ⓒ Ⓓ	
16 Ⓐ Ⓑ Ⓒ Ⓓ	36 Ⓐ Ⓑ Ⓒ Ⓓ	
17 Ⓐ Ⓑ Ⓒ Ⓓ	37 Ⓐ Ⓑ Ⓒ Ⓓ	
18 Ⓐ Ⓑ Ⓒ Ⓓ	38 Ⓐ Ⓑ Ⓒ Ⓓ	
19 Ⓐ Ⓑ Ⓒ Ⓓ	39 Ⓐ Ⓑ Ⓒ Ⓓ	
20 Ⓐ Ⓑ Ⓒ Ⓓ	40 Ⓐ Ⓑ Ⓒ Ⓓ	

ISEE Upper-Level Practice Test 1

Mathematics

Quantitative Reasoning

❖ **37 Questions.**

❖ **Total time for this test: 35 Minutes.**

❖ **Calculators are not allowed at the test.**

Administered *Month Year*

1) Which of the following is NOT a factor of 90?

    A. 9   C. 15

    B. 45   D. 16

2) There are 8 blue marbles, 5 red marbles, and 7 yellow marbles in a box. If Mia randomly selects a marble from the box, what is the probability of selecting a blue or yellow marble?

    A. $\frac{8}{15}$ C. $\frac{3}{4}$

    B. $\frac{7}{15}$ D. $\frac{3}{7}$

3) On a map, the length of the road from City A to City B is measured to be 22 inches. On this map, $\frac{1}{3}$ inch represents an actual distance of 5 miles. What is the actual distance, in miles, from City A to City B along this road?

    A. 48   C. 320

    B. 330   D. 130

4) What is the area of a square whose diagonal is 14?

    A. 28   C. 196

    B. 49   D. 98

5) If Emily left a $34.25 tip on a breakfast that cost $67.50, approximately what percentage was the tip?

    A. 15%   C. 51%

    B. 21%   D. 62%

6) A phone company charges $8 for the first four minutes of a phone call and 40 cents per minute thereafter. If Sofia makes a phone call that lasts 25 minutes, what will be the total cost of the phone call?

A. 14.20   C. 16.40

B. 8.80   D. 16

7) Michelle and Alec can finish a job together in 30 minutes. If Michelle can do the job by herself in 2 hours, how many minutes does it take Alec to finish the job?

A. 80   C. 75

B. 40   D. 145

8) Elise earns $7.20 per hour and worked 20 hours. Bob earns $8.00 per hour. How many hours would Bob need to work to equal James's earnings over 20 hours?

A. 24   C. 14

B. 40   D. 18

9) In the figure, *MN* is 60 cm. How long is ON?

A. 20 cm

B. 35 cm

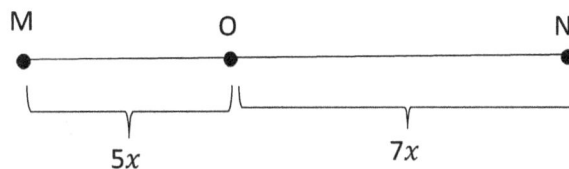

C. 25 cm

D. 28 cm

Use following graph to answer question.

A library has 550 books that include Mathematics, Physics, Chemistry, English and History.

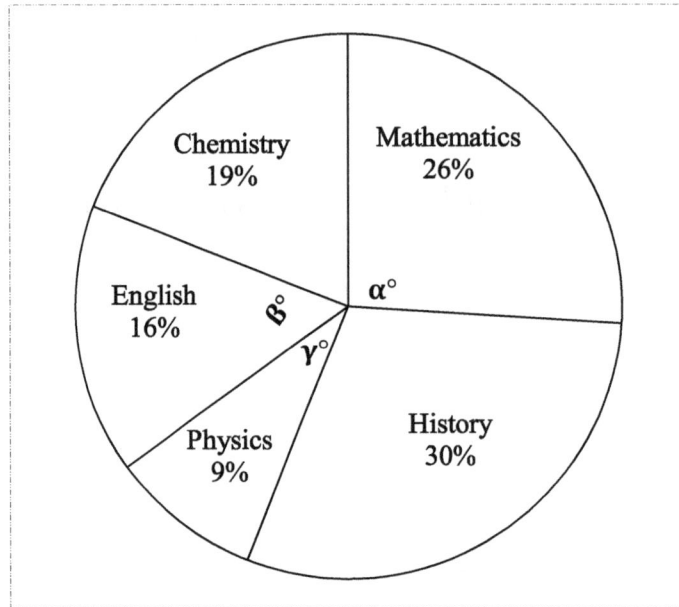

10) What is the product of the number of History and number of English books?

A. 14,220   C. 7,550

B. 14,520   D. 87,520

11) If 150% of a number is 45, then what is the 30% of that number?

A. 9   C. 18

B. 6   D. 15

12) If $7y + 8 < 64$, then y could be equal to?

A. 10.5   C. 9.5

B. 11   D. 7.5

13) The first five terms in a sequence are shown below. What is the seventh term in the sequence?

$$\{4, 7, 12, 19, 28, ...\}$$

A. 52   C. 36

B. 49   D. 42

14) Emily and Daniel have taken the same number of photos on their school trip. Emily has taken 5 times as many as photos as Claire and Daniel has taken 28 more photos than Claire. How many photos has Claire taken?

A. 7   C. 6

B. 5   D. 8

15) What is the equation of the line that passes through (4, –2) and has a slope of 3?

A. $y = 3x - 14$ C. $y = -3x + 12$

B. $y = -3x - 14$ D. $y = 3x + 12$

16) A supermarket's sales increased by 40 percent in the month of April and decreased by 40 percent in the month of May. What is the percent change in the sales of the supermarket over the two-month period?

A. 14% increase   C. 16% decrease

B. No change   D. 4% decrease

17) How many $\frac{1}{6}$ pound paperback books together weigh 14 pounds?

A. 8.4   C. 84

B. 11.4   D. 114

---

18) Find the solution (x, y) to the following system of equations?

$$-2x + 3y = 3$$

$$-4x + 7y = 9$$

A. $(2, 3)$ C. $(3, -3)$

B. $(-3, 2)$ D. $(3, 3)$

19) The distance between cities A and B is approximately 4,350 miles. If Alice drives an average of 75 miles per hour, how many hours will it take Alice to drive from city A to city B?

A. Approximately 58 hours   C. Approximately 68 hours

B. Approximately 87 hours   D. Approximately 61 hours

20) Four fifth of 45 is equal to $\frac{3}{8}$ of what number?

A. 36   C. 56

B. 48   D. 96

21) Sophia purchased a sofa for $493.60. The sofa is regularly priced at $617. What was the percent discount Sophia received on the sofa?

A. 20%   C. 45%

B. 55%   D. 15%

ISEE Upper-Level Subject Test Mathematics

Quantitative Comparisons

Direction: Questions 22 to 37 are Quantitative Comparisons Questions. Using the information provided in each question, compare the quantity in column A to the quantity in Column B. Choose on your answer sheet grid.

A   if the quantity in Column A is greater

B   if the quantity in Column B is greater

C   if the two quantities are equal.

D   if the relationship cannot be determined from the information given.

22) Set A includes odd primes numbers less than 15.

Column A	Column B
The sum of all members in Set A	42

23)

Column A	Column B
1.7%	$\frac{1}{8}$

24)

Column A	Column B
The average of 13, 15, 18, 24, 35	The average of 18, 23, 31, 26

25)

Column A	Column B
$9 + 13(7 - 6)$	$13 + 9(7 - 6)$

26)

Column A	Column B
The number of posts needed for a fence 189 feet long if the posts are placed 13.5 feet apart	17 posts

27) $\frac{x}{64} = \frac{5}{16}$

Column A	Column B
$\dfrac{4}{x}$	$\dfrac{1}{5}$

28) Working at constant rates, machine D makes b rolls of steel in 18 minutes and machine E makes b rolls of steel in 15 minutes ($b > 0$)

Column A	Column B
The number of rolls of steel made by machine D in 1 hours and 30 minutes.	The number of rolls of steel made by machine E in 1 hours.

29) The ratio of boys to girls in a class is 15 to 20.

Column A	Column B
Ratio of girls to the entire class	$\dfrac{1}{7}$

30) There are 9 blue marbles and 5 green marbles in a jar. Two marbles are pulled out in succession without replacing them in the jar.

Column A	**Column B**
The probability that both marbles are blue.	The probability that the first marbles is green, but the second is blue.

31) $\frac{x}{4} = y^2$

Column A	**Column B**
x	y

32) $x = 1$

Column A	**Column B**
$4x^2 - 3x + 15$	$2x^3 + x^2 + 12$

33) $6 > y > -3$

Column A	**Column B**
$\dfrac{y}{6}$	$\dfrac{6}{y}$

34) $\frac{a}{b} = \frac{d}{c}$

Column A	**Column B**
$b - a$	$c - d$

35) A magazine printer consecutively numbered the pages of a magazine, starting with 1 on the first page, 12 on the twelfth page, etc. In numbering the pages, the printer printed a total of 217 digits.

Column A	Column B
The number of pages in the magazine	108

36) A computer priced $159 includes 6% profit.

Column A	Column B
$145	The original cost of the computer

37)

Column A	Column B
The largest number that can be written by rearranging the digits in 386	The largest number that can be written by rearranging the digits in 546

STOP

IF YOU FINISH BEFORE TIME IS CALLED, YOU MAY CHECK YOUR WORK ON THIS SECTION ONLY. DO NOT TURN TO ANY OTHER SECTION IN THE TEST.

ISEE Upper-Level Practice Test 1

Mathematics

Mathematics Achievement

- ❖ **47 Questions.**
- ❖ **Total time for this test: 40 Minutes**.
- ❖ **Calculators are not allowed at the test.**

Administered *Month Year*

1) $9 - 18 \div (9^2 \div 9) =$ ___

    A. 7   C. -5

    B. -7 D. 5

2) How is this number written in scientific notation?

$$0.00004568$$

    A. 4.568×10^5 C. $4,568 \times 10^4$

    B. 4.568×10^{-5} D. $4,567 \times 10^{-4}$

3) A girl 150 cm tall, stands 180 cm from a lamp post at night. Her shadow from the light is 30 cm long. How high is the lamp post?

    A. 500

    B. 1,500

    C. 1,050

    D. 2,100

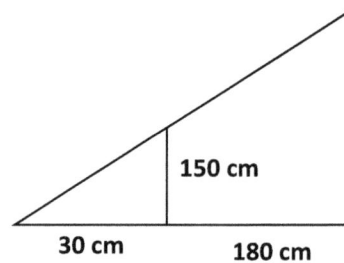

4) Which value of x makes the following inequality true?

$$\frac{5}{28} \le x < 21\%$$

    A. 0.23   C. $\sqrt{0.0121}$

    B. $\frac{13}{67}$ D. $(0.18)^2$

5) $(x + 6)(x + 4) =$

    A. $x^2 + 24x + 10$ C. $x^2 + 10x + 24$

    B. $2x - 48x - 10$ D. $x^2 - 10x + 24$

6) Which of the following graphs represents the compound inequality $-4 \leq 3x -$

7 < 11 ?

A.

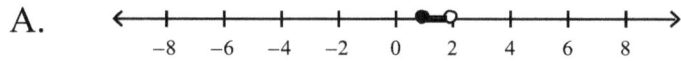

B.

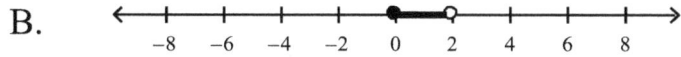

C.

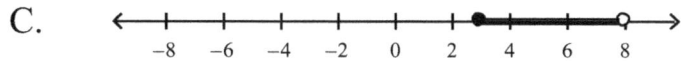

D.

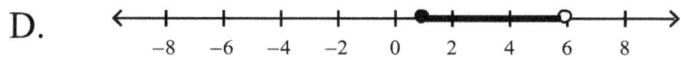

7) Find all values of x for which $8x^2 + 14x + 6 = 0$

A. $-\frac{1}{4}, -1$ C. $3, \frac{3}{4}$

B. $-\frac{3}{4}, -1$ D. $-3, \frac{1}{4}$

8) Emily lives $7\frac{5}{7}$ miles from where she works. When traveling to work, she walks

to a bus stop $\frac{1}{6}$ of the way to catch a bus. How many miles away from her house

is the bus stop?

A. $2\frac{1}{7}$ Miles   C. $2\frac{5}{7}$ Miles

B. $1\frac{5}{14}$ Miles   D. $1\frac{2}{7}$ Miles

9) $|11 - (63 \div |1 - 8|)| = ?$

A. 3   C. 2

B. −2   D. −3

10) The ratio of boys to girls in a school is 2:3. If there are 180 students in a school, how many boys are in the school?

A. 72   C. 46

B. 144   D. 108

11) When an integer is multiplied by itself, they can end in all of the following digits EXCEPT

A. 2, 5   C. 4, 6

B. 1, 4   D. 3, 8

12) The rectangle ABCD on the coordinate grid is translated 5 units down and 2 units to the left.

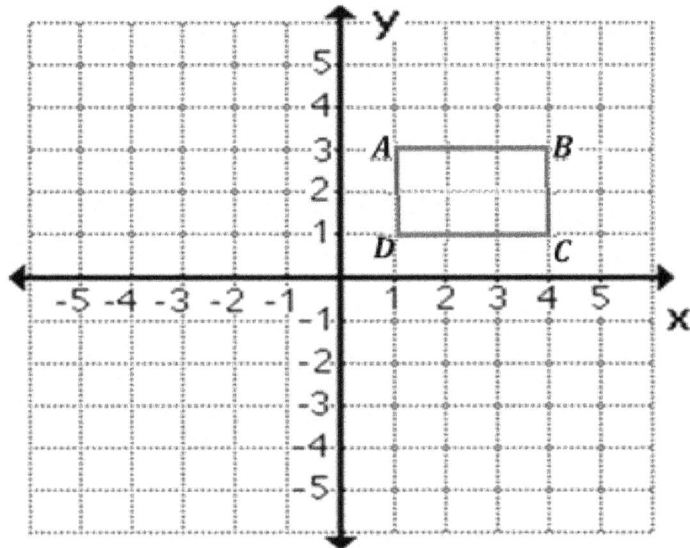

Which of the following describes this transformation?

A. $(x, y) \Rightarrow (x - 5, y - 2)$ C. $(x, y) \Rightarrow (x - 5, y + 2)$

B. $(x, y) \Rightarrow (x - 2, y - 5)$ D. $(x, y) \Rightarrow (x + 2, y - 5)$

13) Which graph shows a non-proportional linear relationship between x and y?

A.

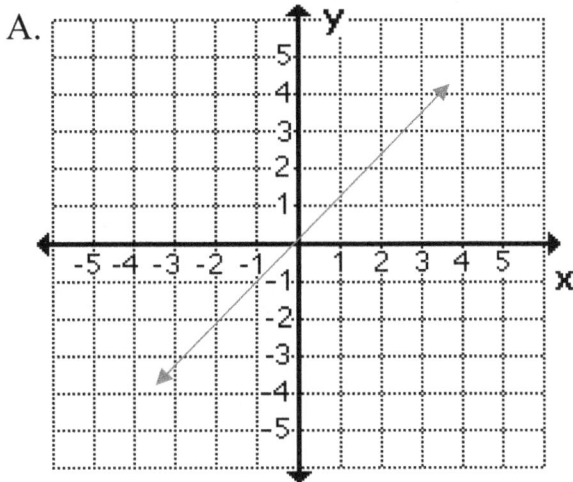

B.

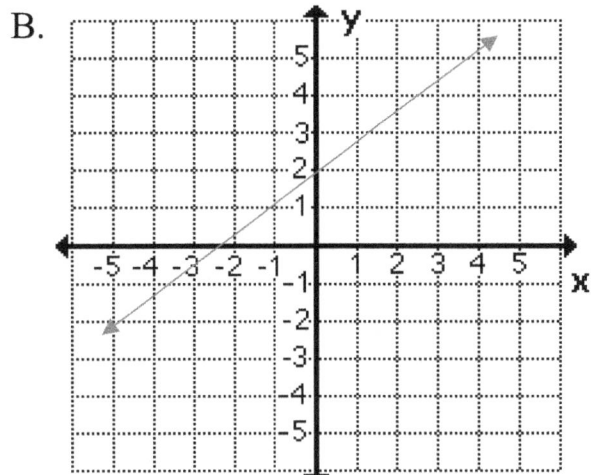

C.

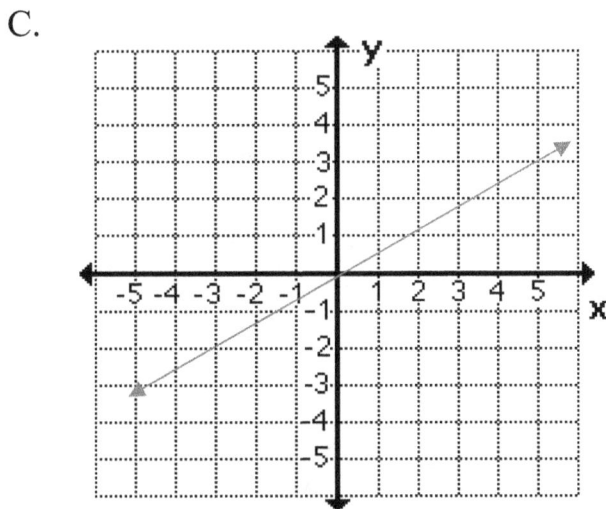

D.

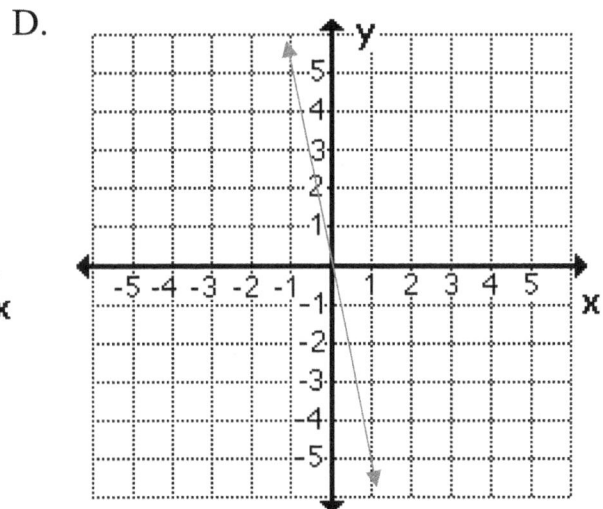

14) Use the diagram below to answer the question.

Given the lengths of the base and diagonal of the rectangle below, what is the length of height h, in terms of s?

A. $3s\sqrt{6}$

B. $2s\sqrt{3}$

C. $24s$

D. $25s$

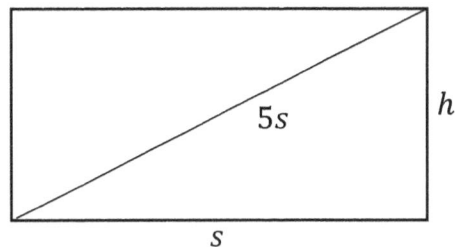

Use the chart below to answer the question.

Color	Number
White	10
Black	15
Beige	35

15) There are also purple marbles in the bag. Which of the following can **NOT** be the probability of randomly selecting a purple marble from the bag?

A. $\frac{1}{6}$

C. $\frac{3}{5}$

B. $\frac{6}{7}$

D. $\frac{2}{9}$

16) If the area of trapezoid is 504 cm, what is the perimeter of the trapezoid?

A. 42 cm

B. 46 cm

C. 92 cm

D. 106 cm

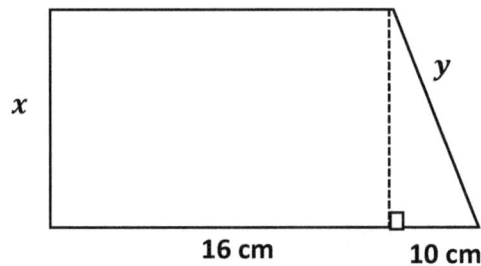

17) If a vehicle is driven 43 miles on Monday, 44 miles on Tuesday, and 36 miles on Wednesday, what is the average number of miles driven each day?

A. 41 Miles

C. 43 Miles

B. 46 Miles

D. 45 Miles

18) Find the area of a rectangle with a length of 116 feet and a width of 47 feet.

A. 5,542 sq. ft

C. 5,245 sq. ft

B. 5,452 sq. ft

D. 5,642 sq. ft

19) $57 \div \frac{1}{4} = ?$

    A. 23.8   C. 148

    B. 42.6   D. 228

20) With an 25% discount, Ella was able to save $17.25 on a dress. What was the

    original price of the dress?

    A. $71   C. $69

    B. $59   D. $57

21) $\frac{8}{65}$ is equals to:

    A. 1.230   C. 0.123

    B. 1.023   D. 0.153

22) If 40% of A is 1,600, what is 12% of A?

    A. 600   C. 480

    B. 360   D. 240

23) If $(7.3 + 2.6 + 8.1) \times x = x$, then what is the value of x?

    A. 0   C. −3

    B. $\frac{1}{8}$ D. −8

24) Two dice are thrown simultaneously, what is the probability of getting a sum of

    3 or 8?

    A. $\frac{1}{18}$ C. $\frac{1}{10}$

    B. $\frac{5}{36}$ D. $\frac{5}{12}$

25) Simplify $\dfrac{\dfrac{1}{2}-\dfrac{x+7}{4}}{\dfrac{x^2}{2}-\dfrac{3}{2}}$

A. $\dfrac{5-x}{x^2+12}$

C. $\dfrac{5-x}{x^2+6}$

B. $\dfrac{5+x}{2x^2+12}$

D. $\dfrac{-5-x}{2x^2-6}$

26) In the following figure, AB is the diameter of the circle. What is the circumference of the circle?

A. 10π

B. 15π

C. 12π

D. 20π

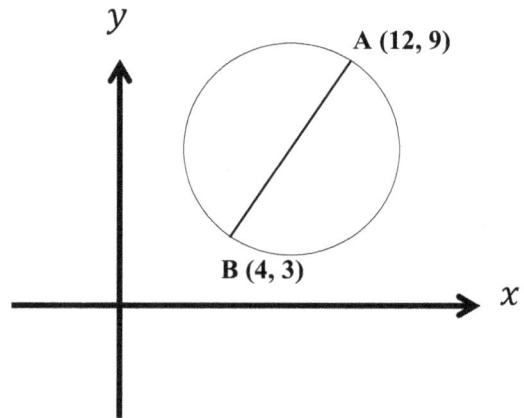

27) What is the value of x in the following equation?

$$3x^2 - 23 = 52$$

A. ± 6

C. ± 5

B. ± 3

D. ± 4

28) A circle has a diameter of 14 inches. What is its approximate area?

A. 156.26

C. 125.55

B. 153.86

D. 136.66

29) If 9 garbage trucks can collect the trash of 81 homes in a day. How many trucks are needed to collect in 270 houses?

A. 46

C. 26

B. 50

D. 30

Use the following table to answer question below.

DANIEL'S BIRD-WATCHING PROJECT	
Day	Number of Raptors Seen
Monday	?
Tuesday	5
Wednesday	16
Thursday	15
Friday	10
Mean	11

30) The above table shows the data Daniel collects while watching birds for one week. How many raptors did Daniel see on Monday?

A. 11

C. 9

B. 7

D. 12

31) $53.28 \div 0.07 =?$

A. 753.14

C. 771.12

B. 761.14

D. 7.721

32) A floppy disk shows 747,205 bytes free and 788,425 bytes used. If you delete a file of size 710,250 bytes and create a new file of size 685,651 bytes, how many free bytes will the floppy disk have?

A. 870,804   C. 678,602

B. 1,158,008   D. 771,804

33) 6 days 2 hours 40 minutes – 4 days 22 hours 55 minutes =?

A. 2 days 1 hours 25 minutes

B. 1 days 3 hours 45 minutes

C. 1 days 3 hours 15 minutes

D. 2 days 1 hours 45 minutes

34) The base of a right triangle is 16 feet, and the interior angles are 45-45-90. What is its area?

A. 128 square feet   C. 32 square feet

B. 118 square feet   D. 64 square feet

35) A circle is inscribed in a square, as shown below. The area of the circle is 16π cm². What is the area of the square?

A. 60 cm²

B. 16 cm²

C. 36 cm²

D. 64 cm²

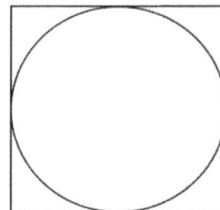

36) Increased by 60%, the number 25 becomes:

    A. 45

    B. 25

    C. 40

    D. 105

37) If $39 + x^{\frac{1}{4}} = 41$, then what is the value of $14 \times x$?

    A. 16   C. 220

    B. 60   D. 224

38) Triangle ABC is graphed on a coordinate grid with vertices at A $(-4, -3)$,

    B $(8, -7)$ and C $(3, 9)$. Triangle ABC is reflected over x axes to create

    triangle A'B'C'.

    Which order pair represents the coordinate of C'?

    A. $(3, 9)$ C. $(-3, 9)$

    B. $(-9, -3)$ D. $(3, -9)$

39) Which set of ordered pairs represents y as a function of x?

    A. $\{(3, -1), (5, 0), (8, -2), (3, 6)\}$

    B. $\{(2, 5), (7, -4), (3, 1), (7, -8)\}$

    C. $\{(15, -9), (-2, 5), (3, 11), (-2, 9)\}$

    D. $\{(6, 1), (-2, 3), (4, 6), (8, -1)\}$

40) Which equation represents the statement "twice the difference between 7 times H and 5 gives 52".

A. $\frac{7H + 2}{5} = 52$

C. $2(7H - 5) = 52$

B. $7(2H + 5) = 52$

D. $5\frac{7H}{2} = 52$

41) David makes a weekly salary of $500 plus 5% commission on his sales. What will his income be for a week in which he makes sales totaling $3,600?

A. $210

C. $580

B. $755

D. $180

42) $8x^2y^6 + 4x^3y^5 - (5x^2y^6 - 4x^3y^5) = $ ___

A. $-2x^3y^5 - 8x^2y^6$

C. $8x^2y^6$

B. $x^2y^6 + 8x^3y^5$

D. $8x^3y^5 + 3x^2y^6$

43) The radius of circle A is three times the radius of circle B. If the circumference of circle A is 36π, what is the area of circle B?

A. 12π

C. 36π

B. 18π

D. 6π

44) If a box contains red and blue balls in ratio of 3: 8 red to blue, how many red balls are there if 80 blue balls are in the box?

A. 15

C. 25

B. 30

D. 40

45) The width of a box is one fourth of its length. The height of the box is one fifth of its width. If the length of the box is 40 cm, what is the volume of the box?

A. 400 cm^3 C. 800 cm^3

B. 480 cm^3 D. 880 cm^3

46) A square measures 8 inches on one side. By how much will the area be decreased if its length is increased by 6 inches and its width decreased by 4 inches.

A. 8 sq. decreased.   C. 12 sq. decreased.

B. 16 sq. decreased.   D. 10 sq. decreased.

47) How many 3 × 3 squares can fit inside a rectangle with a height of 54 and width of 15?

A. 60   C. 95

B. 90   D. 85

STOP

IF YOU FINISH BEFORE TIME IS CALLED, YOU MAY CHECK YOUR WORK ON THIS SECTION ONLY. DO NOT TURN TO ANY OTHER SECTION IN THE TEST.

ISEE Upper-Level Practice Test 2

Mathematics

Quantitative Reasoning

- ❖ **37 Questions.**
- ❖ **Total time for this test: 35 Minutes.**
- ❖ **Calculators are not allowed at the test.**

Administered *Month Year*

1) The area of a circle is less than 25π. Which of the following can be the circumference of the circle?

    A. 25π C. 30π

    B. 10π D. 40π

2) The circle graph below shows all Mr. Green's expenses for last month. If he spent $320 on his car, how much did he spend for his rent?

    A. $640

    B. $720

    C. $680

    D. $520

Mr. Green's monthly expenses

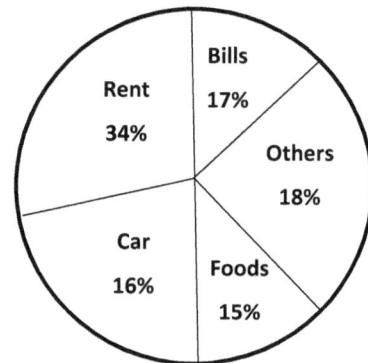

Rent 34%, Bills 17%, Others 18%, Foods 15%, Car 16%

3) Etta drives from her house to work at an average speed of 52 miles per hour and she drives at an average speed of 50 miles per hour when she was returning home. What was her minimum speed on the round trip in miles per hour?

    A. 50   C. 52

    B. 51   D. Cannot be determined.

4) How much greater is the value of $5x + 8$ than the value of $5x - 7$?

    A. 8   C. 56

    B. 15   D. 12

5) Oscar purchased a new hat that was on sale for $11.57. The original price was $22.16. What percentage discount was the sale price?

A. 54.70%   C. 47.78%

B. 40.17%   D. 45.70%

6) If $f(x) = 4x^2 + 16$, what is the smallest possible value of $f(x)$?

A. 12   C. 16

B. −4   D. 4

7) If the sum of the positive integers from 1 to $n + 1$ is 2,758, and the sum of the positive integers from $n + 2$ to $2n$ is 8,453, which of the following represents the sum of the positive integers from 1 to $2n$ inclusive?

A. 10,211   C. 8,611

B. 12,211   D. 11,211

8) Triangle ABC is similar to triangle ADE. What is the length of side EC?

A. 30

B. 15

C. 25

D. 45

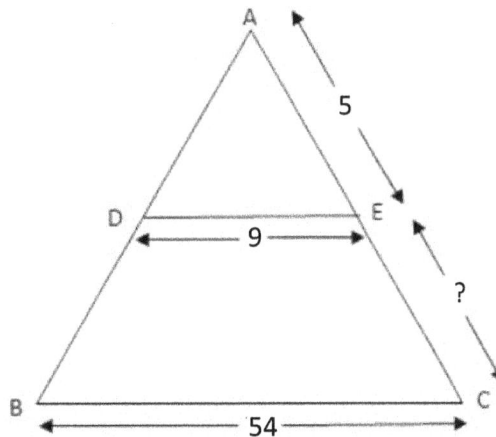

9) Which of the following statements is correct, according to the graph below?

Number of Books Sold in a Bookstore

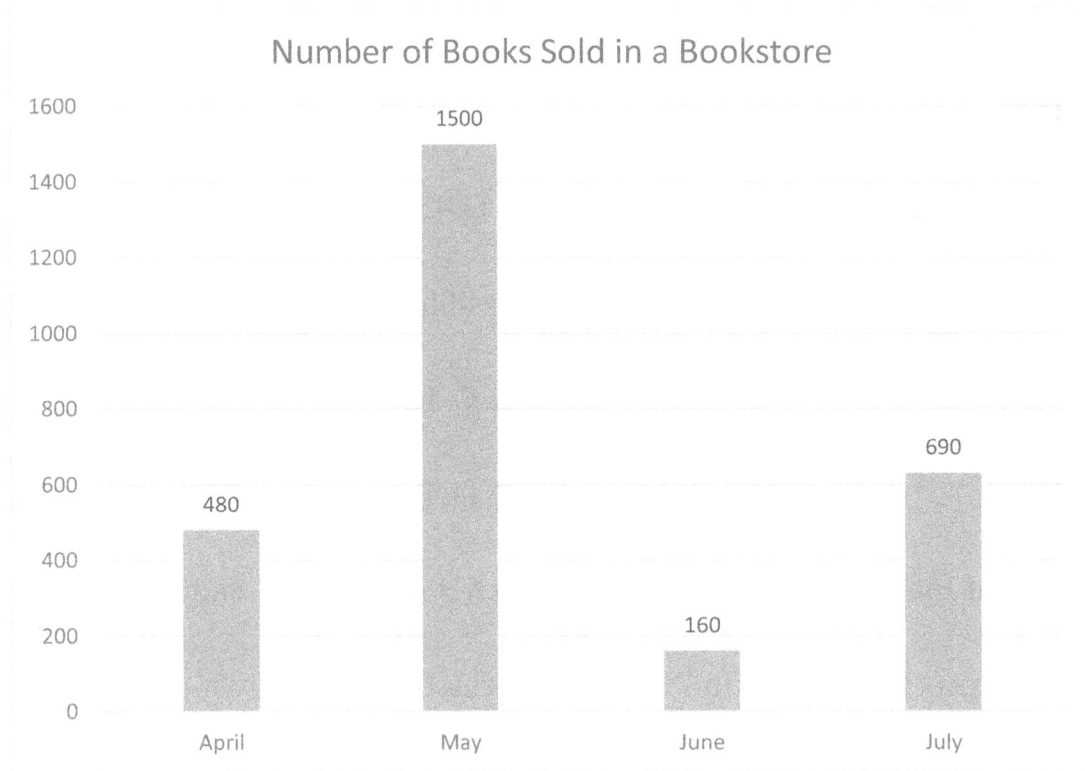

A. Number of books sold in June was greater than one third the number of books sold in July.

B. Number of books sold in July was One third the number of books sold in May.

C. Number of books sold in April was triple the number of books sold in June.

D. Number of books sold in July was equal to the number of books sold in April plus the number of books sold in June.

10) What is the prime factorization of 300?

A. $2 \times 3 \times 5 \times 5$ C. 5×7

B. $2 \times 3 \times 7 \times 7$ D. $2 \times 2 \times 3 \times 5 \times 5$

11) List A consists of the numbers {5, 6, 10, 14, 23}, and list B consists of the numbers {3, 9, 12, 17, 21}.

If the two lists are combined, what is the median of the combined list?

A. 11   C. 9

B. 13   D. 13.5

12) If 14 inches on a map represents an actual distance of 70 feet, then what actual distance does 34 inches on the map represent?

A. 54   C. 64

B. 300   D. 170

13) If Jim adds 40 stamps to his current stamp collection, the total number of stamps will be equal to $\frac{4}{3}$ the current number of stamps. If Jim adds 30% more stamps to the current collection, how many stamps will be in the collection?

A. 460   C. 116

B. 316   D. 156

14) A gas tank can hold 24 gallons when it is $\frac{3}{8}$ full. How many gallons does it contain when it is full?

A. 10   C. 20

B. 64   D. 18.5

15) A basket contains 25 balls and the average weight of each of these balls is 16 g. The four heaviest balls have an average weight of 20 g each. If we remove the six heaviest balls from the basket, what is the average weight of the remaining balls?

A. 12   C. 10.5

B. 16   D. 15.5

16) A bag contains 16 balls: six green, two black, five blue, one brown, one red and one white. If 10 balls are removed from the bag at random, what is the probability that a brown ball has been removed?

A. $\frac{1}{8}$ C. $\frac{5}{8}$

B. $\frac{1}{10}$ D. $\frac{1}{16}$

17) If $3x + 2y = 7$ and $3x - 2y = 5$ then what is the value of $(9x^2 - 4y^2)$?

A. 35   C. 15

B. 25   D. 40

18) What's The ratio of boys and girls in a class is 5:6. If there are 88 students in the class, how many more boys should be enrolled to make the ratio 1:1?

A. 40   C. 18

B. 48   D. 8

19) The area of rectangle ABCD is 243 square inches. If the length of the rectangle is three times the width, what is the perimeter of rectangle ABCD?

A. 72   C. 48

B. 35   D. 86

20) The sum of 6 numbers is greater than 480 and less than 540 Which of the following could be the average (arithmetic mean) of the numbers?

A. 8   C. 65

B. 75   D. 85

21) Which of the following expressions gives the value of z in terms of a, b, and c from the following equation?

$$a = [\frac{3cz}{4b}]^2$$

A. $z = 12ac^2b^2$

B. $z = \frac{4b\sqrt{a}}{3C}$

C. $z = \frac{cb}{\sqrt{3a}}$

D. $z = [\frac{3b}{4ac}]^2$

Quantitative Comparisons

Direction: Questions 22 to 37 are Quantitative Comparisons Questions. Using the information provided in each question, compare the quantity in column A to the quantity in Column B. Choose on your answer sheet grid.

A   if the quantity in Column A is greater

B   if the quantity in Column B is greater

C   if the two quantities are equal.

D   if the relationship cannot be determined from the information given.

22)

Column A	Column B
The average of 15, 32, and 19	21

23)

Column A	Column B
$12 \times 254 \times 17$	$14 \times 254 \times 17$

24) x is an integer:

Column A	Column B
x	$\dfrac{x}{-9}$

25)

Column A	Column B
The greatest value of x in $6\|5x - 9\| = 6$	The greatest value of x in $6\|5x + 9\| = 6$

26) x is an integer:

Column A	Column B
$(x)^3(x)^4$	$(x^4)^3$

27)

Column A	Column B
3^2	$\sqrt[3]{729}$

28)

Column A	Column B
5	$(18)^{\frac{1}{2}}$

29) The selling price of a sport jacket including 30% discount is $210.

Column A	Column B
Original price of the sport jacket	$280

30)

Column A	Column B
$(0.66)^{22}$	$(0.66)^{21}$

31)

Column A	Column B
The probability that	The probability that
event x will occur.	event x will not occur.

32)

Column A	Column B
The probability of rolling a 3 on a die and getting heads on a coin toss.	The probability of rolling an odd number on a die and picking a spade or heart from a deck of 52 cards.

33)

Column A	Column B
$\left(\frac{1}{7}\right)^2$	7^{-2}

34) $4x + 4 > 8x$

Column A	Column B
x	-3

35)

Column A	Column B
$1.45	Sum of four quarter, two nickels, and six pennies

36) x is an even integer, and y is an odd integer, in a certain game an even number is considered greater than an odd number?

Column A	Column B
$(x + y)^2 - y$	$(y)(x - y)$

37) $x^2 - 3x - 6 = 22$

Column A	**Column B**
x	2

STOP

YOU FINISH BEFORE TIME IS CALLED; YOU MAY CHECK YOUR WORK ON THIS

SECTION ONLY. DO NOT TURN TO ANY OTHER SECTION IN THE TEST.

ISEE Upper-Level Practice Test 2

Mathematics

Mathematics Achievement

❖ **47 Questions.**

❖ **Total time for this test: 40 Minutes.**

❖ **Calculators are not allowed at the test.**

Administered *Month Year*

1) $\frac{10}{25}$ is equal to:

   A. 4.0   C. 0.04

   B. 0.40   D. 0.004

2) 13 less than twice a positive integer is 35. What is the integer?

   A. 28   C. 24

   B. 12   D. 48

3) Which of the following points lies on the line $4x + 5y = 17$?

   A. $(2, -1)$ C. $(-1, 5)$

   B. $(-2, 4)$ D. $(3, 1)$

4) An angle is equal to one fourth of its supplement. What is the measure of that angle?

   A. 18   C. 36

   B. 24.5   D. 45

5) 4.8 is what percent of 60?

   A. 1.2   C. 16

   B. 8   D. 2

6) Right triangle ABC has two legs of lengths 15 cm (AB) and 20 cm (AC). What is the length of the third side (BC)?

   A. 18 cm   C. 22 cm

   B. 30 cm   D. 25 cm

7) $\frac{1}{6b^2} + \frac{1}{6b} = \frac{1}{3b^2}$, then $b =$?

A. $-\frac{2}{5}$

C. $-\frac{1}{6}$

B. 1

D. 2

8) If $\frac{|6+x|}{4} \leq 5$, then which of the following is correct?

A. $-14 \leq x \leq 26$

C. $-14 \leq x \leq 14$

B. $-26 \leq x \leq 14$

D. $-26 \leq x \leq 26$

9) The cost, in thousands of dollars, of producing x thousands of textbooks is

$C(x) = x^2 + 3x$. The revenue, also in thousands of dollars, is $R(x) = 18x$. find

the profit or loss if 5 textbooks are produced. (profit = revenue − cost)

A. $25 profit

C. $90 loss

B. $50 profit

D. $40 loss

10) Ella (E) is 9 years older than her friend Ava (A) who is 6 years younger than

her sister Sofia (S). If E, A and S denote their ages, which one of the following

represents the given information?

A. $\begin{cases} E = A + 9 \\ S = A + 6 \end{cases}$

C. $\begin{cases} A = E + 9 \\ A = S + 6 \end{cases}$

B. $\begin{cases} E = A - 9 \\ S = A - 6 \end{cases}$

D. $\begin{cases} A = E - 9 \\ A = S - 6 \end{cases}$

11) Simplify $8x^4y^3(3x^2y)^2 =$

A. 24^8y^4 C. $48x^8y^5$

B. $36x^5y^8$ D. $72x^8y^5$

12) Which is the longest time?

A. 26 hours   C. 1 days and 5 hours

B. 3,990 minutes   D. 101,200 seconds

13) Write 589 in expanded form, using exponents.

A. $(5 \times 10^3) + (8 \times 10^2) + (9 \times 10)$

B. $(5 \times 10^2) + (8 \times 10^1) - 10$

C. $(5 \times 10^2) + (8 \times 10^1) + 9$

D. $(5 \times 10^1) + (8 \times 10^2) + 9$

14) A company pays its writer $6 for every 600 words written. How much will a writer earn for an article with 450 words?

A. $4.6   C. $4.5

B. $ 2.3   D. $45

15) A circular logo is enlarged to fit the lid of a jar. The new diameter is 60% larger than the original. By what percentage has the area of the logo increased?

A. 40%   C. 65%

B. 96%   D. 9%

16) $62.70 \div 0.06 = ?$

A. 10.45   C. 10.045

B. 1,045   D. 104.5

17) A bread recipe calls for $5\frac{3}{4}$ cups of flour. If you only have $2\frac{5}{8}$ cups, how much more flour is needed?

A. 8   C. 3

B. $2\frac{1}{8}$ D. $3\frac{1}{8}$

18) The equation of a line is given as: $y = 4x - 3$. Which of the following points does not lie on the line?

A. $(-1, -7)$ C. $(-2, 10)$

B. $(1, 1)$ D. $(2, 5)$

19) A circle has a diameter of 15 inches. What is its approximate circumference?

A. 62   C. 32

B. 47   D. 42

20) What's the area of the non-shaded part of the following figure?

A. $40\ cm^2$

B. $80\ cm^2$

C. $120\ cm^2$

D. $95\ cm^2$

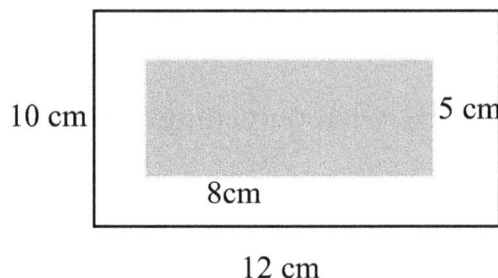

21) What is the area of an isosceles right triangle with hypotenuse that measures 6 cm?

A. 9 cm

C. $3\sqrt{2}$ cm

B. 18 cm

D. 6 cm

22) The drivers at G & G trucking must report the mileage on their trucks each week. The mileage reading of Ed's vehicle was 23,658 at the beginning of one week, and 24,157 at the end of the same week. What was the total number of miles driven by Ed that week?

A. 97 Miles

C. 499 Miles

B. 197 Miles

D. 1,494 Miles

23) What is the maximum value for y if $y = -(x-4)^2 + 9$?

A. -9

C. 5

B. -5

D. 9

24) What is the solution of the following system of equations?

$$\begin{cases} -2x + y = -2 \\ 5x + 4y = 5 \end{cases}$$

A. $(-1, 2)$

C. $(1, 2)$

B. $(1, 0)$

D. $(0, -1)$

25) What is the area of an isosceles right triangle that has one leg that measures 12 cm?

A. 72 cm^2

C. $6\sqrt{2}$ cm^2

B. 14 cm^2

D. 42 cm^2

26) Which of the following is a factor of both $x^2 + 6x - 16$ and $x^2 - 8x + 12$?

    A. $(x - 2)$ C. $(x - 8)$

    B. $(x + 2)$ D. $(x + 8)$

27) If $x + y = 13$, what is the value of $3x + 3y$?

    A. 16   C. 39

    B. 13   D. 36

28) A car uses 15 gallons of gas to travel 480 miles. How many miles per gallon does the car use?

    A. 16 miles per gallon   C. 32 miles per gallon

    B. 24 miles per gallon   D. 10 miles per gallon

29) $(6x + 6)(x + 3) =$

    A. $6x + 24$ C. $6x^2 + 18x + 24$

    B. $6x^2 + 24x + 18$ D. $6x^2 + 18$

30) If $x \blacksquare y = \sqrt{x^2 + 7y}$, what is the value of $6 \blacksquare 4$?

    A. $\sqrt{62}$ C. 10

    B. 8   D. 14

31) There are two equal tanks of water. If $\frac{7}{8}$ of a tank contains 140 liters of water, what is the capacity of the two tanks of water together?

    A. 160   C. 120

    B. 480   D. 320

32) The average weight of 30 girls in a class is 50 kg and the average weight of 20

boys in the same class is 60 kg. What is the average weight of all the 50 students

in that class?

A. 46   C. 52.1

B. 54   D. 42.1

33) If x is 26% percent of 650, what is x?

A. 206   C. 269

B. 169   D. 106

34) $\frac{\begin{array}{r} 16 \text{ hr. } 25 \text{ min.} \\ - 11 \text{ hr. } 47 \text{ min.} \end{array}}{}$

A. 5 hr. 22 min.   C. 15 hr. 32 min.

B. 4 hr. 38 min.   D. 14 hr. 28 min.

35) What is the number of cubic feet of soil needed for a flower box 8 feet long, 6

inches wide, and 9 feet deep?

A. 36 cubic feet   C. $\frac{70}{3}$ cubic feet

B. 18 cubic feet   D. 72 cubic feet

36) Karen is 5 years older than her sister Michelle, and Michelle is 2 years younger

than her brother David. If the sum of their ages is 88, how old is Michelle?

A. 81   C. 18

B. 27   D. 9

37) Mario loaned Jett $620 at a yearly interest rate of 5%. After three years what is the interest owned on this loan?

    A. $1,335   C. $87

    B. $188   D. $93

38) Mason just got hired for on-the-road sales and will travel about 2,400 miles a week during a 54-hour work week. If the time spent traveling is $\frac{4}{9}$ of his week, how many hours a week will he be on the road?

    A. Mason spends about 6 hours of his 54-hour work week on the road.

    B. Mason spends about 12 hours of his 54-hour work week on the road.

    C. Mason spends about 22 hours of his 54-hour work week on the road.

    D. Mason spends about 24 hours of his 54-hour work week on the road.

39) A shirt costing $610 is discounted 32%. After a month, the shirt is discounted another 15%. Which of the following expressions can be used to find the selling price of the shirt?

    A. $(610)\,(0.15)$ C. $(610)(0.15) - (610)(0.32)$

    B. $(610) - 610\,(0.32)$ D. $(610)\,(0.68)\,(0.85)$

40) What is the reciprocal of $\frac{x^5}{22}$?

    A. $\frac{22}{x^5} - 3$ C. $\frac{22}{x^5} + 3$

    B. $\frac{44}{x^5}$ D. $\frac{22}{x^5}$

41) Given that $x = 0.2$ and $y = 5$ what is the value of $3x^2(y + 8)$?

A. 15

C. 5.6

B. 1.56

D. 56

42) Calculate the area of a parallelogram with a base of 7 feet and height of 6.8 feet.

A. 42.6 square feet

C. 47.6 square feet

B. 39.8 square feet

D. 42.8 square feet

43) In a school, the ratio of number of boys to girls is 5:3. If the number of boys is 220, what is the total number of students in the school?

A. 44

C. 244

B. 352

D. 1,044

44) A tree 48 feet tall casts a shadow 16 feet long. Jack is 9 feet tall. How long is Jack's shadow?

A. 3 ft

C. 3.8 ft

B. 2.6 ft

D. 4 ft

45) What is the area of the shaded region if the diameter of the bigger circle is 14 inches and the diameter of the smaller circle is 4 inches?

A. 40 π inch²

B. 45 π inch²

C. 49 π inch²

D. 144 π inch²

46) What is the result of the expression?

$$\begin{vmatrix} 4 & 8 \\ -1 & -1 \\ -5 & 8 \end{vmatrix} + \begin{vmatrix} 0 & -4 \\ 4 & -2 \\ 4 & -6 \end{vmatrix} =$$

A. $\begin{vmatrix} 4 & -4 \\ 3 & -3 \\ 1 & 2 \end{vmatrix}$

C. $\begin{vmatrix} 4 & 4 \\ 3 & -3 \\ -1 & 2 \end{vmatrix}$

B. $\begin{vmatrix} 4 & -4 \\ -3 & -2 \\ -1 & -4 \end{vmatrix}$

D. $\begin{vmatrix} 4 & -4 \\ -3 & 3 \\ -12 & -1 \end{vmatrix}$

47) How many square feet of tile is needed for 9 feet by 9 feet room?

A. 18 square feet

C. 81 square feet

B. 36 square feet

D. 27 square fee

STOP

IF YOU FINISH BEFORE TIME IS CALLED, YOU MAY CHECK YOUR WORK ON THIS SECTION ONLY. DO NOT TURN TO ANY OTHER SECTION IN THE TEST.

Chapter 13 : Answers and Explanations

ISEE Upper-Level Practice Tests

Answer Key

✴ Now, it's time to review your results to see where you went wrong and what areas you need to improve!

ISEE Upper-Level Practice Test 1 - Mathematics

Quantitative Reasoning						Mathematics Achievement							
1	D	16	C	31	D	1	A	16	C	31	B	46	A
2	C	17	C	32	A	2	B	17	A	32	D	47	B
3	B	18	D	33	D	3	C	18	B	33	B		
4	D	19	A	34	D	4	B	19	D	34	A		
5	C	20	D	35	A	5	C	20	C	35	D		
6	C	21	A	36	B	6	D	21	C	36	C		
7	B	22	B	37	A	7	B	22	C	37	D		
8	D	23	B			8	D	23	A	38	D		
9	B	24	B			9	C	24	B	39	D		
10	B	25	C			10	A	25	D	40	C		
11	A	26	B			11	D	26	A	41	C		
12	D	27	C			12	B	27	C	42	D		
13	A	28	A			13	B	28	B	43	C		
14	A	29	A			14	B	29	D	44	B		
15	A	30	A			15	D	30	C	45	C		

Answers and Explanations

ISEE Upper-Level Practice Tests

ISEE Upper-Level Practice Test 2 - Mathematics

Quantitative Reasoning

1	B	16	D	31	D
2	C	17	A	32	B
3	D	18	D	33	C
4	B	19	A	34	D
5	C	20	D	35	A
6	C	21	B	36	A
7	D	22	A	37	D
8	C	23	B		
9	C	24	D		
10	D	25	A		
11	A	26	D		
12	D	27	C		
13	D	28	A		
14	B	29	A		
15	B	30	B		

Mathematics Achievement

1	B	16	B	31	D	46	C
2	C	17	D	32	B	47	C
3	D	18	C	33	B		
4	C	19	B	34	B		
5	B	20	B	35	A		
6	D	21	A	36	B		
7	B	22	C	37	D		
8	B	23	D	38	D		
9	B	24	B	39	D		
10	A	25	A	40	D		
11	D	26	A	41	B		
12	B	27	C	42	C		
13	C	28	C	43	B		
14	C	29	B	44	A		
15	B	30	B	45	B		

Score Your Test

ISEE Upper-Level scores are broken down by four sections: Verbal Reasoning, Reading Comprehension, Quantitative Reasoning, and Mathematics Achievement. A sum of the ALL sections is also reported. The Essay section is scored separately. For the Upper-Level ISEE, the score range is 760 to 940, the lowest possible score a student can earn is 760 and the highest score is 940 for each section. A student receives 1 point for every correct answer. There is no penalty for wrong or skipped questions.

The total scaled score for an Upper-Level ISEE test is the sum of the scores for all sections. A student will also receive a percentile score of between 1-99% that compares that student's test scores with those of other test takers of same grade and gender from the past 3 years. When a student receives her/his score, the percentile score is also be broken down into a stanine and the stanines are ranging from 1–9. Most schools accept students with scores of 5–9. The ideal candidate has scores of 6 or higher.

The following charts provide an estimate of students ISEE percentile rankings for the practice tests, compared against other students taking these tests. Keep in mind that these percentiles are estimates only, and your actual ISEE percentile will depend on the specific group of students taking the exam in your year.

Percentile Rank	Stanine
1 – 3	1
4 – 10	2
11 - 22	3
23 – 39	4
40 – 59	5
60 – 76	6
77 - 88	7
89 – 95	8
96 – 99	9

ISEE Upper-Level Quantitative Reasoning Percentiles

Grade Applying to	25th Percentile	50th Percentile	75th Percentile
9th	850	880	895
10th	855	885	900
11th	860	890	905
12th	864	892	908

Use the next table to convert ISEE Upper-level raw score to scaled score for application to 9th - 12th grade.

ISEE Upper-Level Subject Test Mathematics

ISEE Upper-Level Scaled Scores

Raw Score	Quantitative Reasoning Report Range		Mathematics Achievement Report Range		Raw Score	Quantitative Reasoning Report Range		Mathematics Achievement Report Range	
0	760	760	760	760	26	900	885	885	865
1	770	765	770	765	27	905	890	885	865
2	780	770	780	770	28	910	895	890	870
3	790	775	790	775	29	910	900	890	870
4	800	780	800	780	30	915	905	895	875
5	810	785	810	785	31	920	910	895	875
6	820	790	820	790	32	925	915	900	880
7	825	795	825	795	33	930	920	900	880
8	830	800	830	800	34	930	925	905	885
9	835	805	835	805	35	935	930	905	885
10	840	810	840	810	36	935	935	910	890
11	845	815	845	815	37	940	940	910	890
12	850	820	850	820	38			915	895
13	855	825	855	825	39			920	900
14	860	830	855	830	40			925	905
15	865	835	860	835	41			925	910
16	870	840	860	840	42			930	915
17	875	845	865	840	43			930	920
18	880	845	865	845	44			935	925
19	880	850	870	845	45			935	930
20	885	855	870	850	46			940	935
21	885	860	875	850	47			940	940
22	890	865	875	855					
23	890	870	875	855					
24	895	875	880	860					
25	895	880	880	860					

Answers and Explanations

ISEE - Upper-Level

Practice Tests 1: Quantitative Reasoning

1) Answer: D.

A factor must divide evenly into its multiple. 14 cannot be a factor of 90 because 90 divided by 16 = 5.63

2) Answer: C.

Probability = $\frac{number\ of\ desired\ outcomes}{number\ of\ total\ outcomes}$

In this case, a desired outcome is selecting either a blue or a yellow marble. Combine the number of blue and yellow marbles: $8 + 7 = 15$ and divide this by the total number of marbles: $8 + 7 + 5 = 20$. The probability is $\frac{15}{20} = \frac{3}{4}$.

3) Answer: B.

The distance on the map is proportional to the actual distance between the two cities. Use the information to set up a proportion and then solve for the unknown number of actual miles: $\frac{5\ miles}{\frac{1}{3}\ inches} = \frac{x\ miles}{18\ inches}$

Cross multiply and simplify to solve for the x:

$\frac{5 \times 22}{\frac{1}{3}} = x\ miles \rightarrow \frac{110}{\frac{1}{3}} = 110 \times 3 = 330\ miles$

4) Answer: D.

The diagonal of the square is 14. Let x be the side.

Use Pythagorean Theorem: $a^2 + b^2 = c^2$

$x^2 + x^2 = 14^2 \Rightarrow 2x^2 = 14^2 \Rightarrow 2x^2 = 196 \Rightarrow x^2 = 98 \Rightarrow x = \sqrt{98}$

The area of the square is: $\sqrt{98} \times \sqrt{98} = 98$

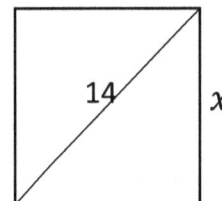

5) Answer: C.

To find what percent A is of B, divide A by B, then multiply that number by 100%:

$34.25 \div 67.50 = 0.5074 \times 100\% = 50.74\%$

This is approximately 51 %.

6) Answer: C.

The total cost of the phone call can be represented by the equation: $TC = \$8.00 + \$0.4(x - 4)$, where x is the duration of the call after the first four minutes. In this case, $x = 25$. Substitute the known values into the equation and solve:

$TC = \$8.00 + \$0.4 \times (25 - 4)$

$TC = \$8.00 + \$8.4 \Rightarrow TC = \$16.4$

7) Answer: B.

Let b be the amount of time Alec can do the job, then,

$\frac{1}{a} + \frac{1}{b} = \frac{1}{30} \rightarrow \frac{1}{120} + \frac{1}{b} = \frac{1}{30} \rightarrow \frac{1}{b} = \frac{1}{30} - \frac{1}{120} = \frac{3}{120} = \frac{1}{40}$

Then: $b = 40$ minutes

8) Answer: D.

Begin by calculating Elise total earnings after 20 hours:

20 hours $\times$ \$7.20 per hour = \$144

Next, divide this total by Bob's hourly rate to find the number of hours Bob would need to work: $\$144 \div \$8.00$ per hour = 18 hours

9) Answer: B.

The length of MN is equal to: $5x + 7x = 12x$

Then: $12x = 60 \rightarrow x = \frac{60}{12} = 5$

The length of ON is equal to: $7x = 7 \times 5 = 35$ cm.

10) Answer: B.

Number of History book: $0.3 \times 550 = 165$

Number of English books: $0.16 \times 550 = 88$

Product of number of History and number of English books: $165 \times 88 = 14,520$

11) Answer: A.

First, find the number.

Let x be the number. Write the equation and solve for x.

150 % of a number is 45, then: $1.5 \times x = 45 \rightarrow x = 45 \div 1.5 = 30$

30% of 30 is: $0.30 \times 30 = 9$

12) Answer: D.

$7y + 8 < 64 \rightarrow 7y < 64 - 8 \rightarrow 7y < 56 \rightarrow y < 8$

Only choice D (7.5) is less than 8.

13) Answer: A.

Begin by examining the sequence to find the pattern. The difference between 4 and 7 is 3; moving from 7 to 12 requires 5 to be added; moving from 12 to 19 requires 7 to be added; moving from 19 to 28 requires 9 to be added. The pattern emerges here — adding by consecutive odd integers. The 6th term is equal to $28 + 11 = 39$, and the 7th term is equal to $39 + 13 = 52$.

14) Answer: A.

Write equations based on the information provided in the question: Emily = Daniel,

Emily = 5 Claire,   Daniel = 28 + Claire

$\rightarrow$5 Claire = 28 + Claire $\rightarrow$5 Claire – Claire = 28

4 Claire = 28,   Claire = 7

15) Answer: A.

The general slope-intercept form of the equation of a line is $y = mx + b$, where m is the slope and b is the y-intercept.

By substitution of the given point and given slope:$-2 = (3)(4) + b$

So, $b = -2 - 12 = -14$ and the required equation is $y = 3x - 14$

16) Answer: C.

Let's choose $100 for the sales of the supermarket. If the sales increases by 40 percent in April, the final number of sales at the end of April will be $100 + (40\%) \times (\$100) = \$140$.

If sales then decreased by 40 percent in May, the final number of sales at the end of May will be $\$140 – (\$140) \times (40\%) = 140 – 56 = \$84$

The final sales of $84 is 84% of the original price of $100.

Therefore, the sales decreased by 16% overall.

17) Answer: C.

If each book weighs $\frac{1}{6}$ pound, then 1 pound = 6 books. To find the number of books in

14 pounds, simply multiply this 6 by 14: $14 \times 6 = 84$

18) **Answer: D.**

Multiplying each side of $-2x + 3y = 3$ by -2 gives $4x - 6y = -6$.

Adding each side of $4x - 6y = -6$ to the corresponding side of $-4x + 7y = 9$ gives $y = 3$.

Finally, substituting 3 for y in $-4x + 7y = 9$ gives $-4x + 7(3) = 9$ or $x = 3$.

19) **Answer: A.**

The time it takes to drive from city A to city B is: $\frac{4,350}{75} = 58$

20) **Answer: D.**

Let x be the number. Write the equation and solve for x.

$\frac{4}{5} \times 45 = \frac{3}{8}x \rightarrow \frac{4 \times 45}{5} = \frac{3x}{8}$, use cross multiplication to solve for x.

$32 \times 45 = 3x \times 5 \Rightarrow 1,440 = 15x \Rightarrow x = 96$

21) **Answer: A.**

The question is this: 493.60 is what percent of 617?

Use percent formula:

Part $= \frac{percent}{100} \times$ whole

$493.60 = \frac{percent}{100} \times 617 \rightarrow 493.60 = \frac{percent \times 617}{100} \rightarrow 49,360 = percent \times 617$

Then, Percent $= \frac{49,360}{617} = 80$

493.60 is 80% of 617. Therefore, the discount is: $100\% - 80\% = 20\%$

22) **Answer: B.**

Recall that the first and only even prime number is 2. The other prime numbers are: 3, 5, 7, 11, 13, 17, etc. They are all odd numbers less than 15 are 3, 5, 7,11, 13 so the sum of members in Set A is $3 + 5 + 7 + 11 + 13 = 39$. So, the correct answer is B.

23) **Answer: B.**

Column A: $1.7\% = 0.017$

Column B: $\frac{1}{8} = 0.125$

0.125 is greater than 0.017.

24) Answer: B.

Column A: $\frac{13+15+18+24+35}{5} = \frac{105}{5} = 21$

Column B: $\frac{18 + 23 + 31 + 26}{4} = \frac{98}{4} = 24.5$

25) Answer: C.

Column A: $9 + 13(7 - 6) = 22$

Column B: $13 + 9(7 - 6) = 22$

26) Answer: B.

The posts are placed 13.5 feet apart. Since a post is needed at the very beginning as well as at the end, B requires 17 posts. $189 \div 13.5 = 14 \Longrightarrow 14 + 3 = 17$

27) Answer: C.

First find the value of x.

$\frac{x}{64} = \frac{5}{16} \rightarrow 16x = 5 \times 64 = 320 \rightarrow x = \frac{320}{16} = 20$

Column A: $\frac{4}{x} = \frac{4}{20} = \frac{1}{5}$

28) Answer: A.

First convert hours to minutes. 1 hours 30 minutes $= 1 \times 60 + 30 = 90$ minutes.

Machine D makes b rolls of steel in 18 minutes. So, it makes 5 sets of b in 90 minutes. $90 \div 18 = 5$ sets of b.

Machine E operates for 1 hours, making b rolls per 15 minutes. So, it makes a total of $4b$ rolls. Therefore, machine D makes more rolls, and Column A is greater.

29) Answer: A.

The ratio of boys to girls in a class is 15 to 20. Therefore, ratio of girls to the entire class is 15 out of 35. $\frac{15}{35} = \frac{3}{7} > \frac{1}{7}$

30) Answer: A.

Let us calculate each probability individually:

That probability that the first marble is blue $= \frac{9}{14}$

The probability that the second marble is blue $= \frac{8}{13}$

Column A: The probability that both marbles are blue: $\frac{9}{14} \times \frac{8}{13} = \frac{72}{182} = \frac{36}{91}$

The probability that the marble is green $= \frac{5}{14}$

The probability that the second marble is blue $= \frac{4}{13}$

Column B: The probability that the first marble is green, but the second is blue

$= \frac{5}{14} \times \frac{4}{13} = \frac{20}{182} = \frac{10}{91}$

Column A is greater. $\frac{36}{91} > \frac{10}{91}$.

31) Answer: D.

First, solve the expression for x. $\frac{x}{4} = y^2 \rightarrow x = 4y^2$

Plug in different values for y and find the values of x.

Let's choose $y = 0 \rightarrow x = 4y^2 \rightarrow x = 4(0)^2 = 0$

The values in Column A and B are equal.

Now, let's choose $y = 1 \rightarrow x = 4y^2 \rightarrow x = 4(1)^2 = 4$

Column A is greater. So, the relationship cannot be determined from the information given.

32) Answer: A.

Column A: $4x^2 - 3x + 15 = 4(1)^2 - 3(1) + 15 = 4 - 3 + 15 = 16$

Column B: $2x^3 + x^2 + 12 = 2(1)^3 + (1)^2 + 12 = 2 + 1 + 12 = 15$

33) Answer: D.

$6 > y > -3$

Let's choose some values for y. $y = -1$

Column A: $\frac{y}{6} = \frac{-1}{6}$

Column B: $\frac{6}{y} = \frac{6}{-1} = -6$

In this case, column A is bigger. $y = 1$

Column A: $\frac{y}{6} = \frac{1}{6}$

Column B: $\frac{6}{y} = \frac{6}{1} = 6$

In this case, Column B is bigger. So, the relationship cannot be determined from the information given.

34) **Answer: D.**

$\frac{a}{b} = \frac{d}{c}$ Here there are two equal fractions.

Let's choose some values for these variables. $\frac{1}{3} = \frac{3}{9}$

In this case, Column A is 2 (3 − 1) and Column B is 6 (9 − 3). Since we can change the positions of these variables (for example put 3 for a and 9 for b), here relationship cannot be determined from the information given.

35) **Answer: A.**

First, let's find the number of digits when the printer prints 108 pages.

If there are 2 digits in each page and the printer prints 108 pages, then, there will be 216 digits. $108 \times 2 = 216$

However, we know that pages 1–9 have only one digit each, so we must subtract 9 from this total: $216 − 9 = 205$. We also know that the number 100th to 108th have three digits not two. So, we must add $9 \times 1 = 9$ digit (we counted two digits before) to this total: $205 + 9 = 213$.

It is given that 217 digits were printed, and we know that 108 pages results in 213 digits total, so there must be 109 total pages in the magazine. Column A is greater.

36) **Answer: B.**

The computer priced $159 includes 6% profit. Let x be the original cost of the computer. Then: $x + 6\% \ of \ x = 159 \rightarrow x + 0.06x = 159 \rightarrow 1.06x = 159 \rightarrow x = \frac{159}{1.06} = 150$

Column B is bigger.

37) **Answer: A.**

Column A: The largest number that can be written by rearranging the digits in 386 = 863

Column B: The largest number that can be written by rearranging the digits in 546 = 654

Answers and Explanations

ISEE - Upper-Level

Practice Tests 1: Mathematics Achievement

1) Answer: A.

Simplify: $9 - 18 \div (9^2 \div 9) = 9 - 18 \div (81 \div 9) = 9 - 18 \div (9) = 9 - 2 = 7$

2) Answer: B.

$0.00004568 = \dfrac{4.568}{100,000} \Rightarrow 4.568 \times 10^{-5}$

3) Answer: C.

Write the proportion and solve for missing side.

$\dfrac{\text{Smaller triangle height}}{\text{Smaller triangle base}} = \dfrac{\text{Bigger triangle height}}{\text{Bigger triangle base}} \Rightarrow \dfrac{30cm}{150cm} = \dfrac{30+180cm}{x} \Rightarrow x = 1,050 \text{ cm}$

4) Answer: B.

$\dfrac{5}{28} = 0.178$ and $21\% = 0.21$ therefore x should be between 0.178 and 0.21

Only choice B ($\dfrac{13}{67} = 0.194$) is between 0.178 and 0.21

5) Answer: C.

Use FOIL (First, Out, In, Last) method.

$(x + 6)(x + 4) = x^2 + 4x + 6x + 24 = x^2 + 10x + 24$

6) Answer: D.

Solve for x.

$-4 \le 3x - 7 < 11 \Rightarrow$ (add 7 all sides) $-4 + 7 \le 3x - 7 + 7 < 11 + 7 \Rightarrow 3 \le 3x < 18 \Rightarrow$ (divide all sides by 3) $1 \le x < 6$

x is between 1 and 6. Choice D represent this inequality.

7) Answer: B.

First, divide all values on both sides of the equation by 2.

Then: $4x^2 + 7x + 3 = 0$

Use quadratic formula: $ax^2 + bx + c = 0$

$x_{1,2} = \dfrac{-b \pm \sqrt{b^2 - 4ac}}{2a}$

$4x^2 + 7x + 3 = 0 \Rightarrow$ then: $a = 4, b = 7$ and $c = 3$

$$x = \frac{-7 + \sqrt{7^2 - 4 \times 4 \times 3}}{2 \times 4} = -\frac{6}{8} = -\frac{3}{4}$$

$$x = \frac{-7 - \sqrt{7^2 - 4 \times 4 \times 3}}{2 \times 4} = -\frac{8}{8} = -1$$

8) Answer: D.

$\frac{1}{6}$ of the distance $7\frac{5}{7}$ miles is: $\frac{1}{6} \times 7\frac{5}{7} = \frac{1}{6} \times \frac{54}{7} = \frac{9}{7}$

Converting $\frac{9}{7}$ to a mixed number gives: $\frac{9}{7} = 1\frac{2}{7}$

9) Answer: C.

Simplify: $|11 - (63 \div |1 - 8|)| = |11 - (63 \div |-7|)| = |11 - (63 \div 7)| =$

$|11 - 9| = |2| = 2$

10) Answer: A.

The ratio of boy to girls is 2:3. Therefore, there are 2 boys out of 5 students. To find the answer, first divide the total number of students by 5, then multiply the result by 2.

$180 \div 5 = 36 \Rightarrow 36 \times 2 = 72$

11) Answer: D.

All Integers must end in one of the following digits:

0 when multiplied by itself ends in 0.

1 when multiplied by itself ends in 1.

2 when multiplied by itself ends in 4.

3 when multiplied by itself ends in 9.

4 when multiplied by itself ends in 6.

5 when multiplied by itself ends in 5.

6 when multiplied by itself ends in 6.

7 when multiplied by itself ends in 9.

8 when multiplied by itself ends in 4.

9 when multiplied by itself ends in 1 .

Number 3,7,8 is not in the results.

12) Answer: B.

Translated 5 units down and 2 units to the left means:

$(x, y) \Rightarrow (x - 2, y - 5)$

13) Answer: B.

A linear equation is a relationship between two variables, x and y, and can be written in the form of $y = mx + b$

A non-proportional linear relationship takes on the form $y = mx + b$, where $b \neq 0$ and its graph is a line that does not cross through the origin.

Only in graph B, the line does not pass through the origin.

14) Answer: B.

Use Pythagorean theorem:

$a^2 + b^2 = c^2 \rightarrow s^2 + h^2 = (5s)^2 \rightarrow s^2 + h^2 = 25s^2$

Subtracting s^2 from both sides gives: $h^2 = 24s^2$

Square roots of both sides: $h = \sqrt{24s^2} = \sqrt{4 \times 3 \times s^2} = 2s\sqrt{3}$

15) Answer: D.

Let x be the number of purple marbles. Let's review the choices provided:

A. $\frac{1}{6}$, if the probability of choosing a purple marble is one out of ten, then:

$Probability = \frac{number\ of\ desired\ outcomes}{number\ of\ total\ outcomes} = \frac{x}{10+15+35+x} = \frac{1}{6}$

Use cross multiplication and solve for x. $6x = 60 + x \rightarrow 5x = 60 \rightarrow x = 12$

Since number of purple marbles can be 6, then, choice be the probability of randomly selecting a purple marble from the bag.

Use same method for other choices.

B. $\frac{6}{7}$; $\frac{x}{10+15+35+x} = \frac{6}{7} \rightarrow 7x = 360 + 6x \rightarrow x = 360 \rightarrow x = 360$

C. $\frac{3}{5}$; $\frac{x}{10+15+35+x} = \frac{3}{5} \rightarrow 5x = 180 + 3x \rightarrow 2x = 180 \rightarrow x = 90$

D. $\frac{2}{9}$; $\frac{x}{10+15+35+x} = \frac{2}{9} \rightarrow 9x = 120 + 2x \rightarrow 7x = 120 \rightarrow x = 17.14$ (Number of

purple marbles cannot be a decimal).

16) Answer: C.

The area of the trapezoid is:

$$Area = \frac{1}{2}h(b_1 + b_2) = \frac{1}{2}(x)(26 + 16) = 504$$

$$\rightarrow 21x = 504 \rightarrow x = 24$$

$$y = \sqrt{10^2 + 24^2} = \sqrt{100 + 576} = \sqrt{676} = 26$$

The perimeter of the trapezoid is: $26 + 16 + 24 + 26 = 92$

17) Answer: A.

$$average = \frac{sum}{total} = \frac{43 + 44 + 36}{3} = \frac{123}{3} = 41$$

18) Answer: B.

Area of a rectangle = width × height

Area = 116 × 47 = 5,452

19) Answer: D.

$$57 \div \frac{1}{4} = 57 \times 4 = 228$$

20) Answer: C.

Let x be the original price of the dress. Then: 25% of x = 17.25

$$\frac{25}{100}x = 17.25$$

$$x = \frac{100 \times 17.25}{25} = 69$$

21) Answer: C.

$$\frac{8}{65} = 0.123$$

22) Answer: C.

40% of A is 1,600 Then: $0.4A = 1,600 \rightarrow A = \frac{1,600}{0.4} = 4,000$

12% of 4,000 is: $0.12 \times 4,000 = 480$

23) Answer: A.

$$(7.3 + 2.6 + 8.1) \times x = x$$

$$18x = x; \text{ Then } x = 0$$

24) Answer: B.

For sum of 3: (1 & 2) and (2 & 1), therefore we have 2 options.

For sum of 8: (2 & 6) and (6 & 2), (4 & 4), we have 3 options.

To get a sum of 3 or 8 for two dice: $2 + 3 = 5$

Since, we have $6 \times 6 = 36$ total number of options, the probability of getting a sum of

3 or 8 is 5 out of 36 or $\dfrac{5}{36}$

25) Answer: D.

Simplify:

$$\frac{\frac{1}{2} - \frac{x+7}{4}}{\frac{x^2}{2} - \frac{3}{2}} = \frac{\frac{2}{4} - \frac{x+7}{4}}{\frac{x^2 - 3}{2}} = \frac{\frac{2-x-7}{4}}{\frac{x^2-3}{2}} = \frac{-x-5}{4} \times \frac{2}{x^2-3}$$

Then: $\dfrac{2(-x-5)}{4(x^2-3)} = \dfrac{-x-5}{2(x^2-3)} = \dfrac{-x-5}{2x^2-6}$

26) Answer: A.

The distance of A to B on the coordinate plane is: $\sqrt{(x_1 - x_2)^2 + (y_1 - y_2)^2} =$

$\sqrt{(12-4)^2 + (9-3)^2} = \sqrt{8^2 + 6^2} = \sqrt{64+36} = \sqrt{100} = 10$

The diameter of the circle is 10 and the radius of the circle is 5. Then: the

circumference of the circle is: $2\pi r = 2\pi(5) = 10\pi$

27) Answer: C.

$3x^2 - 23 = 52 \rightarrow 3x^2 = 75$

$x^2 = 25 \rightarrow x = \pm5$

28) Answer: B.

Diameter = 14, then: Radius = 7

Area of a circle = $\pi r^2 \Rightarrow A = 3.14(7)^2 = 153.86$

29) Answer: D.

Write a proportion and solve. $\dfrac{9}{81} = \dfrac{x}{270} \rightarrow x = \dfrac{9 \times 270}{81} = 30$

30) Answer: C.

The mean of the data is 11. Then:

$\dfrac{x+5+16+15+10}{5} = 11 \rightarrow x + 46 = 55 \rightarrow x = 55 - 46 = 9$

31) Answer: B.

$53.28 \div 0.07 = 761.14$

32) Answer: D.

The difference of the file added, and the file deleted is:

$710,250 - 685,651 = 24,599$

$747,205 + 24,599 = 771,804$

33) Answer: B.

6 days 2 hours 40 minutes – 4 days 22 hours 55 minutes = 1 days and 3 hours and 45 min

34) Answer: A.

Formula of triangle area $= \frac{1}{2}$ (base $\times$ height)

Since the angles are 45-45-90, then this is an isosceles triangle, meaning that the base and height of the triangle are equal.

Triangle area $= \frac{1}{2}$ (base $\times$ height) $= \frac{1}{2}$ (16 $\times$ 16) $= 128$

35) Answer: D.

The area of the circle is 25π cm^2, then, its diameter is 8cm.

$area\ of\ a\ circle = \pi r^2 = 16\pi \rightarrow r^2 = 16 \rightarrow r = 4$

Radius of the circle is 5 and diameter is twice of it, 8.

One side of the square equals to the diameter of the circle. Then:

$Area\ of\ square = side \times side = 8 \times 8 = 64$

36) Answer: C.

60% of 25 = 15; 25 + 15 = 40

37) Answer: D.

$x^{\frac{1}{4}}$ equals to the root of x. Then: $39 + x^{\frac{1}{4}} = 41 \rightarrow 39 + \sqrt[4]{x} = 41 \rightarrow \sqrt[4]{x} = 2 \rightarrow x = 16$

$x = 16$ and $14 \times x$ equals: $14 \times 16 = 224$

38) Answer: D.

When a point is reflected over x axes, the (y) coordinate of that point changes to $(-y)$ while its x coordinate remains the same.

C (3, 9) $\rightarrow$ C' (3, -9)

39) Answer: D.

A set of ordered pairs represents y as a function of x if:

$x_1 = x_2 \rightarrow y_1 = y_2$

In choice A: $(3, -1)$ and $(3, 6)$ are ordered pairs with same x and different y, therefore y isn't a function of x.

In choice B: $(7, -4)$ and $(7, -8)$ are ordered pairs with same x and different y, therefore y isn't a function of x.

In choice C: $(-2, 5)$ and $(-2, 9)$ are ordered pairs with same x and different y, therefore y isn't a function of x.

40) Answer: C.

Only choice C represents the statement "twice the difference between 7 times H and 5 gives 52". $2(7H - 5) = 52$

41) Answer: C.

David's weekly salary is $500 plus 5% of $3,600. Then: $5\% \ of \ 3,600 =$

$0.05 \times 3,600 = 180$

$500 + 180 = 580$

42) Answer: D.

$8x^2y^6 + 4x^3y^5 - (5x^2y^6 - 4x^3y^5) =$

$8x^2y^6 + 4x^3y^5 - 5x^2y^6 + 4x^3y^5 = 8x^3y^5 + 3\ x^2y^6$

43) Answer: C.

Let P be circumference of circle A, then; $2\pi r_A = 36\pi \rightarrow r_A = 18$

$r_A = 3r_B \rightarrow r_B = \frac{18}{3} = 6 \rightarrow$ Area of circle B is; $\pi r_B^2 = 36\pi$

44) Answer: B.

Write a proportion and solve. $\frac{3}{8} = \frac{x}{80}$

Use cross multiplication: $8x = 240 \rightarrow x = 30$

45) Answer: C.

If the length of the box is 40, then the width of the box is one fourth of it, 10, and the height of the box is 2 (one fifth of the width). The volume of the box is:

$V = length \times width \times height = (40)(10)(2) = 800$

46) Answer: A.

The area of the square is 64 square inches. *Area of square = side × side =*

$8 \times 8 = 64$

The length of the square is increased by 6 inches and its width decreased by 4 inches.

Then, its area equals:

Area of rectangle= *width × length* $= (8 + 6) \times (8 - 4) = 56$

The area of the square will be decreased by 8 square inches. $64 - 56 = 8$

47) Answer: B.

Number of squares equal to: $\dfrac{54 \times 15}{3 \times 3} = 18 \times 5 = 90$

Answers and Explanations
ISEE - Upper-Level
Practice Tests 2: Quantitative Reasoning

1) Answer: B.

Area of the circle is less than 25 π. Use the formula of areas of circles.

Area $= \pi r^2 \Rightarrow 25\,\pi > \pi r^2 \Rightarrow 25 > r^2 \Rightarrow r < 5$

Radius of the circle is less than 5. Let's put 5 for the radius. Now, use the circumference formula: Circumference $= 2\pi r = 2\pi\,(5) = 10\,\pi$

Since the radius of the circle is less than 5. Then, the circumference of the circle must be less than 10 π. Only choice B is less than 10π

2) Answer: C.

Let x be all expenses, then $\frac{16}{100}x = \$320 \rightarrow x = \frac{100 \times \$320}{16} = 2{,}000$

He spent for his rent: $\frac{34}{100} \times \$2{,}000 = \$680$

3) Answer: D.

There is not enough information to determine the answer to the question. An average speed represents a distance divided by time and it does not provide information about the speed at specific time. Etta could drove exactly 52 miles per hour from start to finish, or she could drive 50 miles per hour for half of distance and 51 miles per hour for the other half.

4) Answer: B.

$(5x + 8) - (5x - 7) = 5x - 5x + 8 + 7 = 15$

5) Answer: C.

The percentage discount is the reduction in price divided by the original price. The difference between original price and sale price is: $\$22.16 - \$11.57 = \$10.59$

The percentage discount is this difference divided by the original price:

$\$10.59 \div \$22.16 \cong 0.4778$

Convert the decimal to a percentage by multiplying by 100%:

$0.4778 \times 100\% = 47.78\%$

6) Answer: C.

The smallest possible value of $f(x)$ will occur when $x = 0$. Since x^2 is always positive, any positive or negative value of x will make the value of $f(x)$ greater than 16. Substitute 0 for x and evaluate the expression: $f(0) = 4(0)^2 + 16 = 16$

7) Answer: D.

There are 2 sets of values, one set from 1 to $n + 1$, and the other set from $n + 2$ to $2n$. Since the second set begins immediately after the first set, the two sets can be combined. The sum of the positive integers from 1 to $2n$ inclusive is equal to the sum of the positive integers from 1 to $n + 1$ plus the sum of the positive integers from $n + 2$ to $2n$: $2{,}758 + 8{,}453 = 11{,}211$

8) Answer: C.

If two triangles are similar, then the ratios of corresponding sides are equal.

$\frac{AC}{AE} = \frac{BC}{DE} = \frac{54}{9} = 6 \Rightarrow \frac{AC}{AE} = 6$

This ratio can be used to find the length of AC:

$AC = 6 \times AE \Rightarrow AC = 6 \times 5 \Rightarrow AC = 30$

The length of AE is given as 5 and we now know the length of AC is 30, therefore:

$EC = AC - AE \Rightarrow EC = 30 - 5 \Rightarrow EC = 25$

9) Answer: C.

Let's review the choices provided:

A. number of books sold in June is: 160

One third the number of books sold in July is: $\frac{690}{3} = 230 \rightarrow 160 < 230$

B. number of books sold in May is: 1,500

One third the number of books sold in May is: $\frac{1{,}500}{3} = 500 \rightarrow 500 \neq 690$

C. Number of books sold in April is: 480

Number of books sold in June is: $160 \rightarrow \frac{480}{160} = 3$

D. $480 + 160 = 640 < 690$

10) **Answer: D.**

Find the value of each choice:

A. $2 \times 3 \times 5 \times 5 = 150$

B. $2 \times 3 \times 7 \times 7 = 294$

C. $5 \times 7 = 35$

D. $2 \times 2 \times 3 \times 5 \times 5 = 300$

11) **Answer: A.**

The median of a set of data is the value located in the middle of the data set. Combine the 2 sets provided, and organize them in ascending order:

$\{3, 5, 6, 9, 10, 12, 14, 17, 21, 23\}$

Since there are an even number of items in the resulting list, the median is the average of the two middle numbers.

Median $= (10 + 12) \div 2 = 11$

12) **Answer: D.**

Write a proportion and solve.

$\frac{14 in}{70 feet} = \frac{34 in}{x} \rightarrow x = \frac{70 \times 34}{14} = 170 \ feet$

13) **Answer: D.**

Let x be the number of current stamps in the collection. Then:

$\frac{4}{3}x - x = 40 \rightarrow \frac{1}{3}x = 40 \rightarrow x = 120$

30% more of 120 is: $120 + 0.30 \times 120 = 120 + 36 = 156$

14) **Answer: B.**

Let x be number of gallons the tank can hold when it is full. Then:

$\frac{3}{8}x = 24 \rightarrow x = \frac{8}{3} \times 24 = 64$

15) **Answer: B.**

Recall that the formula for the average is:

Average $= \frac{sum \ of \ data}{number \ of \ data}$

First, compute the total weight of all balls in the basket:

$$16\,g = \frac{total\ weight}{25\ balls}$$

$16 \times 25 =$ total weight $= 400$ g

Next, find the total weight of the 4 heaviest balls:

$$20\,g = \frac{total\ weight}{4\ marbles}$$

20 g $\times 4 =$ total weight $= 80$ g

The total weight of six heaviest balls is 80 g. Then, the total weight of the remaining 16 balls is 320 g. 400 g – 80 g = 320 g.

The average weight of the remaining balls:

$$Average = \frac{320\ g}{20\ marbles} = 16\ g\ per\ ball$$

16) Answer: D.

If 12 balls are removed from the bag at random, there will be one ball in the bag. The probability of choosing a brown ball is 1 out of 16. Therefore, the probability of not choosing a brown ball is 10 out of 16 and the probability of having not a brown ball after removing 10 balls is the same.

17) Answer: A.

$(9x^2 - 4y^2) = (3x - 2y)(3x + 2y)$

Then: $(9x^2 - 4y^2) = 7 \times 5 = 35$

18) Answer: D.

The ratio of boy to girls is $5:6$. Therefore, there are 5 boys out of 11 students. To find the answer, first divide the total number of students by 11, then multiply the result by 5.

$88 \div 11 = 8 \Rightarrow 8 \times 5 = 40$

There are 40 boys and 48 (88 – 40) girls. So, 8 more boys should be enrolled to make the ratio 1:1

19) Answer: A.

The formula for the area of a rectangle is: Area = Width × Length

It is given that L = 3W and that A = 243. Substitute the given values into our equation and solve for W:

$243 = w \times 3w$

$243 = 3w^2 \Rightarrow w^2 = 81 \Rightarrow w = 9$

It is given that $L = 3W$, therefore, $L = 3 \times 8 = 27$

The perimeter of a rectangle is: $2L + 2W$

Perimeter $= 2 \times 27 + 2 \times 9$,   Perimeter $= 72$

20) Answer: D.

The sum of 6 numbers is greater than 480 and less than 540. Then, the average of the 6 numbers must be greater than 80 and less than 90.

$\frac{480}{6} < x < \frac{540}{6} \Rightarrow 80 < x < 90$

The only choice that is between 80 and 90 is 85.

21) Answer: B.

In order to solve for the variable b, first take square roots on both sides:

$\sqrt{a} = \frac{3cz}{4b}$, then multiply both sides by b: $4b\sqrt{a} = 3cz$

Now, divide both sides by c: $Z = \frac{4b\sqrt{a}}{3c}$

22) Answer: A.

The average is the sum of all terms divided by the number of terms.

$15 + 32 + 19 = 66$

$66 \div 3 = 22$

This is greater than 21.

23) Answer: B.

Since both columns have 254 as a factor, we can ignore that number.

$12 \times 17 = 204$ $; 14 \times 17 = 238$; Column B is greater

24) Answer: D.

Since x is an integer and can be positive and negative, then the relationship cannot be determined from the information given. Let's choose some values for x.

$x = 1$, then the value in column A is greater. $1 > \frac{1}{-9}$

Let's choose a negative value for x.

$x = -1$, then the value in column B is greater. $-1 < \frac{-1}{-9} \rightarrow -1 < \frac{1}{9}$

25) Answer: A.

First, find the values of x in both columns.

Column A: $6|5x - 9| = 6 \rightarrow |5x - 9| = 1$

$5x - 9$ can be 1 or -1.

$5x - 9 = 1 \rightarrow 5x = 10 \rightarrow x = 2$

$5x - 9 = -1 \rightarrow 5x = 8 \rightarrow x = \frac{8}{5}$

Column B: $6|5x + 9| = 6 \rightarrow |5x + 9| = 1$

$5x + 9$ can be 1 or -1.

$5x + 9 = 1 \rightarrow 5x = -8 \rightarrow x = -\frac{8}{5}$

$5x + 9 = -1 \rightarrow 5x = -10 \rightarrow x = -2$

The greatest value of x in column A is 2 and the greatest value of x in columns B is $-\frac{8}{5}$.

26) Answer: D.

Simplify both columns.

Column A: $(x)^3(x)^4 = x^7$

Column B: $(x^3)^4 = x^{12}$

Column A evaluates to x^7 and Column B evaluates to x^{12}. In the case where $x = 0$, the two columns will be equal, but if $x = 2$, the two columns will not be equal. Consequently, the relationship cannot be determined.

27) Answer: C.

Column A: $3^2 = 9$

Column B: $\sqrt[3]{729} = 9$ (recall that $9^3 = 729$)

28) Answer: A.

A number raised to the exponent $(\frac{1}{2})$ is the same thing as evaluating the square root of the number. Therefore: $(18)^{\frac{1}{2}} = \sqrt{18}$

Since $\sqrt{25}$ is greater than $\sqrt{18}$, column A ($\sqrt{25} = 5$) is greater than $\sqrt{18}$.

29) **Answer: A.**

Let x be the original price of the sport jacket. The selling price of a sport jacket including 30% discount is $210. Then:

$$x - 0.30x = 210 \rightarrow 0.70x = 210 \rightarrow x = \frac{210}{0.70} = 300$$

The original price of the jacket is $300 which is greater than column B ($280).

30) **Answer: B.**

The value of x has to be less than 22, which is less than 22. Column B is greater.

Recall that when the positive powers of numbers between 0 and 1 increases, the value of the number decrease. For example: $(0.7)^2 > (0.7)^3 \rightarrow 0.49 > 0.343$

So, $(0.66)^{33} > (0.66)^{34}$

31) **Answer: D.**

The probability that an event will occur + the probability that that event will NOT occur must equal 1. Since we don't have any numerical information about the probability, it is possible that the probability that event x occurs is 25%, 50% or any other percent. The probability that event x will not occur will always be 100% minus the probability that event x does occur. Because both columns can exhibit a range of values, the relationship cannot be determined.

32) **Answer: B.**

Because of the word "and" the events described in each column must be calculated separately and then multiplied:

For column A: Probability of rolling a 3: $\frac{1}{6}$

Probability of getting heads: $\frac{1}{2}$

$\frac{1}{6} \times \frac{1}{2} = \frac{1}{12}$

For column B: Probability of an odd number: $\frac{3}{6} = \frac{1}{2}$

Probability of getting a spade or heard: $\frac{26}{52} = \frac{1}{2}$

$\frac{1}{2} \times \frac{1}{2} = \frac{1}{4}$

Since $\frac{1}{4}$ is a larger number than $\frac{1}{12}$, Colum B is greater

33) Answer: C.

To raise a quantity to a negative power, invert the numerator and denominator, and then raise the base to the indicated power. Therefore:

$(\frac{7}{1})^{-2} = (\frac{1}{7})^2$; The Columns are the same value.

34) Answer: D.

First, simplify the inequality:

$4x + 4 > 8x \rightarrow 4 > 4x \rightarrow \frac{4}{4} > x \rightarrow 1 > x$

Since x is less than 1, and x can be 0 (greater than -1) or -3 (less than -1), the relationship cannot be determined.

35) Answer: A.

Sum of four quarter, two nickels, and six pennies is: $4(\$0.25) + 2(\$0.05) + 6(\$0.01) = \$1.16$

36) Answer: A.

Let's consider the properties of odd and even integers:

Odd +/− Odd = Even   Odd × Odd = Odd

Even+/−Even= Even   Even × Even = Even

Odd +/− Even = Odd   Odd × Even = Even

For <u>Column A</u>: $(x + y)^2 - y$

(even + odd) ^{2} − odd

(odd)^2 − odd $\Rightarrow$ (odd)(odd) − odd $\Rightarrow$ odd − odd $\Rightarrow$ even

Now let's review the columns. For <u>column B</u>: $(y)(x - y)$

(odd) (even − odd) $\Rightarrow$ (odd) (odd) $\Rightarrow$ odd

Since an even number is considered greater according to the problem statement, the answer is A.

37) Answer: D.

Factor the expression if possible. Begin by moving all terms to one side before factoring:

$x^2 - 3x - 6 = 22 \Rightarrow x^2 - 3x - 28 = 0$

To factor this quadratic, find two numbers that multiply to -48 and sum to -2:

$(x - 7)(x + 4) = 0$

Set each expression in parentheses equal to 0 and solve:

$x - 7 = 0 \Rightarrow x = 7$

$x + 4 = 0 \Rightarrow x = -4$

Quadratic equations can have TWO possible solutions. Since one of these is greater than 0 and one of them is less than 0, we cannot determine the relationship between the columns.

Answers and Explanations

ISEE - Upper-Level

Practice Tests 2: Mathematics Achievement

1) Answer: B.

$$\frac{10}{25} = 0.4$$

2) Answer: C.

Let x be the integer. Then: $2x - 13 = 35$

Add 13 both sides: $2x = 48 \Rightarrow$ Divide both sides by 2: $x = 24$

3) Answer: D.

Plug in each pair of numbers in the equation. The answer should be 11.

A. $(2, -2)$: $4(2) + 5(-2) = -2$ No!

B. $(-2, 4)$: $4(-2) + 5(4) = 12$ No!

C. $(-1, 5)$: $4(-1) + 5(5) = 21$ No!

D. $(3, 1)$: $4(3) + 5(1) = 17$ Yes!

4) Answer: C.

The sum of supplement angles is 180. Let x be that angle. Therefore, $x + 4x = 180$

$5x = 180$, divide both sides by 5: $x = 36$

5) Answer: B.

$x\% \ 60 = 4.8$

$$\frac{x}{100} \ 60 = 4.8 \rightarrow x = \frac{4.8 \times 100}{60} = 8$$

6) Answer: D.

Use Pythagorean Theorem: $a^2 + b^2 = c^2$

$15^2 + 20^2 = c^2 \Rightarrow 625 = c^2 \Rightarrow c = 25$

7) Answer: B.

Subtract $\frac{1}{6b}$ and $\frac{1}{3b^2}$ from both sides of the equation. Then:

$$\frac{1}{6b^2} + \frac{1}{6b} = \frac{1}{3b^2} \rightarrow \frac{1}{6b^2} - \frac{1}{3b^2} = -\frac{1}{6b}$$

Multiply both numerator and denominator of the fraction $\frac{1}{b^2}$ by 2.

Then: $\frac{1}{6b^2} - \frac{2}{6b^2} = -\frac{1}{6b}$

Simplify the first side of the equation: $-\frac{1}{6b^2} = -\frac{1}{6b}$

Use cross multiplication method: $6b = 6b^2 \rightarrow 6 = 6b \rightarrow b = 1$

8) Answer: B.

First, multiply both sides of inequality by 4. Then:

$\frac{|6+x|}{4} \leq 5 \rightarrow |6 + x| \leq 20 \rightarrow -20 \leq 6 + x \leq 20$

Since $6 + x$ can be positive or negative, then:

$6 + x \leq 20 \ or \ 6 + x \geq -20$

Then: $x \leq 14 \ or \ x \geq -26$

Choice B is correct.

9) Answer: B.

Plug in the value of $x = 5$ into both equations. Then:

$C(x) = x^2 + 3x = (5)^2 + 3(5) = 25 + 15 = 40$

$R(x) = 18x = 18 \times 5 = 90$

$90 - 40 = 50$

So, the profit is $50.

10) Answer: A.

From choices provided, only choice D is correct.

$E = A + 9$

$A = S - 6 \Rightarrow S = A + 6$

11) Answer: D.

Simplify. $8x^4y^3(3x^2y)^2 = 8x^4y^3(9x^4y^2) = 72x^8y^5$

12) Answer: B.

26 hours = 93,600 seconds

3,990 minutes = 239,400 seconds

1 days and 5 hours = 29 hours = 104,400 seconds

13) Answer: C.

Let's review the choices provided:

A. $(5 \times 10^3) + (8 \times 10^2) + (9 \times 10) = 5,000 + 800 + 90 = 5,890$

B. $(5 \times 10^2) + (8 \times 10^1) - 10 = 500 + 80 - 10 = 570$

C. $(5 \times 10^2) + (8 \times 10^1) + 9 = 500 + 80 + 9 = 589$

D. $(5 \times 10^1) + (8 \times 10^2) + 95 = 50 + 800 + 9 = 859$

Only choice C equals to 859.

14) Answer: C.

$\frac{6}{600} = \frac{x}{450} \Rightarrow x = \frac{6 \times 450}{600} = 4.5$

15) Answer: B.

Area of a circle equals: $A = \pi r^2$

The new diameter is 60% larger than the original then the new radius is 40% larger than the original (a diameter is twice of a radius).

30% larger than r is $1.4r$. Then, the area of larger circle is:

$A = \pi r^2 = \pi(1.4r)^2 = \pi(1.96r^2) = 1.96\pi r^2$

$1.96\pi r^2$ is 96% bigger than πr^2.

16) Answer: B.

$62.7 \div 0.06 = 1,045$

17) Answer: D.

$5\frac{3}{4} - 2\frac{5}{8} = 5\frac{6}{8} - 2\frac{5}{8} = \frac{46}{8} - \frac{21}{8} = \frac{25}{8} = 3\frac{1}{8}$

18) Answer: C.

Let's review the choices provided. Put the values of x and y in the equation.

A. $(-1, -7)$ $\Rightarrow x = -1 \Rightarrow y = -7$ This is true!

B. $(1, 1)$ $\Rightarrow x = 1 \Rightarrow y = 1$ This is true!

C. $(-2, 10)$ $\Rightarrow x = -2 \Rightarrow y = -11 \neq 10$ This is not true!

D. $(2, 5)$ $\Rightarrow x = 2 \Rightarrow y = 5$ This is true!

19) Answer: B.

$C = 2\pi r = \pi d \rightarrow C = \pi \times 15 = 15\pi$

$\pi = 3.14 \rightarrow C = 15\pi = 47.1 = 47$

20) Answer: B.

The area of the non-shaded region is equal to the area of the bigger rectangle subtracted by the area of smaller rectangle.

Area of the bigger rectangle = $12 \times 10 = 120$

Area of the smaller rectangle = $8 \times 5 = 40$

Area of the non-shaded region = $120 - 40 = 80$

21) Answer: A.

First draw an isosceles triangle. Remember that two sides of the triangle are equal.

Let put a for the legs. Then: Use Pythagorean theorem to find the value of a:

Isosceles right triangle

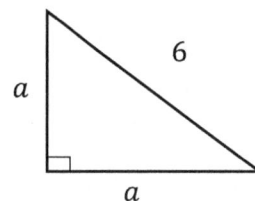

$a^2 + b^2 = c^2 \rightarrow a^2 + a^2 = 6^2$

Simplify: $2a^2 = 36 \rightarrow a^2 = 18 \rightarrow a = \sqrt{18}$

$a = \sqrt{18} \Rightarrow$ area of the triangle is $= \frac{1}{2}\left(\sqrt{18} \times \sqrt{18}\right) = \frac{1}{2} \times 18 = 9 \ cm^2$

22) Answer: C.

To find total number of miles driven by Ed that week, you only need to subtract 23,658 from 24,157. $24,157 - 23,658 = 499$

23) Answer: D.

To find the maximum value of y, the expression $(x - 4)^2$ must be equal to 0.

Because it has a negative sign. Since $x - 4$ is to the power of 4, it cannot be negative.

To get 0 for the expression $(x - 4)^2$, x must be 4.

Plug in 3 for x in the equation: $y = -(x - 4)^2 + 9 \rightarrow y = -(4 - 4)^2 + 9 = 9 \Rightarrow$

The maximum value of y is 8.

24) Answer: B.

$\begin{cases} -2x + y = -2 \\ 5x + 4y = 5 \end{cases} \Rightarrow$ Multiplication (–4) in first equation $\Rightarrow \begin{cases} 8x - 4y = 8 \\ 5x + 4y = 5 \end{cases}$

Add two equations together $\Rightarrow 13x = 13 \Rightarrow x = 1$ then: $y = 0$

25) Answer: A.

First draw an isosceles triangle. Remember that two sides of the triangle are equal.

Isosceles right triangle

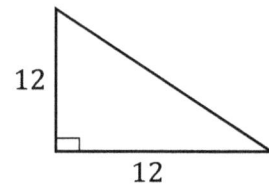

Let put a for the legs. Then:

$a = 12 \Rightarrow$ area of the triangle is $=$

$\frac{1}{2}(12 \times 12) = \frac{144}{2} = 72\ cm^2$

26) Answer: A.

Factor each trinomial $x^2 + 6x - 16$ and $x^2 - 8x + 12$

$x^2 + 6x - 16 \Rightarrow (x - 2)(x + 8)$

$x^2 - 8x + 12 \Rightarrow (x - 6)(x - 2)$

The common factor of both expressions is $(x - 2)$.

27) Answer: C.

$x + y = 13$

Then: $3x + 3y = 3(x + y) = 3 \times 13 = 39$

28) Answer: C.

$\frac{480}{15} = 32$

29) Answer: B.

Use FOIL (First, Out, In, Last)

$(6x + 6)(x + 3) = 6x^2 + 18x + 6x + 18 = 6x^2 + 24x + 18$

30) Answer: B.

Plug in the values of x and y in the equation:

$6 \blacksquare 4 = \sqrt{6^2 + 7(4)} = \sqrt{36 + 28} = \sqrt{64} = 8$

31) Answer: D.

Let x be the capacity of one tank. Then, $\frac{7}{8}x = 140 \rightarrow x = \frac{140 \times 8}{7} = 160$ Liters

The amount of water in two tanks is equal to: $2 \times 160 = 320$ Liters.

32) Answer: B.

$$\text{Average} = \frac{\text{sum of terms}}{\text{number of terms}}$$

The sum of the weight of all girls is: $30 \times 50 = 1,500 \ kg$

The sum of the weight of all boys is: $20 \times 60 = 1,200 \ kg$

The sum of the weight of all students is: $1,500 + 1,200 = 2,700 \ kg$

$$\text{Average} = \frac{2,700}{50} = 54$$

33) Answer: B.

$$\frac{26}{100} \times 650 = x \Rightarrow x = 169$$

34) Answer: B.

$$\begin{aligned} & 16 \text{ hr. } 25 \text{ min.} \\ - \ & 11 \text{ hr. } 47 \text{ min.} \\ \hline & 4 \text{ hr } and \ 38 \text{min} \end{aligned}$$

35) Answer: A.

First, convert all measurement to foot.

One foot is 12 inches. Then: 6 inches $= \frac{6}{12} = \frac{1}{2}$ feet

The volume flower box is length $\times$ width $\times$ height $= 8 \times \frac{1}{2} \times 9 = 36$ cubic feet.

36) Answer: B.

Let's write equations based on the information provided:

Michelle = Karen – 5 $\Rightarrow$ Karen= Michelle + 5

Michelle = David – 2 $\Rightarrow$ David = Michelle + 2

Karen + Michelle + David = 88

Now, replace the ages of Karen and David by Michelle. Then:

Michelle + 5 + Michelle + Michelle + 2 = 88

3Michelle + 7 = 88 $\Rightarrow$ 3 Michelle = 88 – 7

3 Michelle = 81 $\Rightarrow$ Michelle = 27

37) Answer: D.

Use interest rate formula:

$Interest = principal \times rate \times time = 620 \times 0.05 \times 3 = 93$

38) Answer: D.

Mason travels $\frac{4}{9}$ of 54 hours. $\frac{4}{9} \times 54 = 24$

Mason will be on the road for 24 hours.

39) Answer: D.

To find the discount, multiply the number by (100% – rate of discount).

Therefore, for the first discount we get: $(610) (100\% - 32\%) = (610) (0.68)$

For the next 15 % discount: $(610) (0.68) (0.85)$

40) Answer: D.

$\frac{x^5}{22} \Rightarrow$ reciprocal is: $\frac{22}{x^5}$

41) Answer: B.

Plug in the values of x and y in the expression:

$3x^2(y + 8) = 3(0.2)^2(5 + 8) = 3 (0.04)(13) = 1.56$

42) Answer: C.

$A = bh \Rightarrow A = 7 \times 6.8 = 47.6$

43) Answer: B.

The ratio of boys to girls is 5:3. Therefore, there are 5 boys out of 8 students. To find the answer, first divide the number of boys by 5, then multiply the result by 8.

$220 \div 5 = 44 \Rightarrow 44 \times 8 = 352$

44) Answer: A.

Write a proportion and solve for the missing number.

$\frac{48}{16} = \frac{9}{x} \rightarrow 48x = 9 \times 16 = 144$

$48x = 144 \rightarrow x = \frac{144}{48} = 3$

45) Answer: B.

To find the area of the shaded region subtract smaller circle from bigger circle.

$S_{\text{bigger}} - S_{\text{smaller}} = \pi(r_{\text{bigger}})^2 - \pi(r_{\text{smaller}})^2 \Rightarrow S_{\text{bigger}} - S_{\text{smaller}} =$

$\pi(7)^2 - \pi(2)^2 \Rightarrow 49\pi - 4\pi = 45\pi$

46) Answer: C.

To add two matrices, first we need to find corresponding members from each matrix.

$$\begin{vmatrix} 4 & 8 \\ -1 & -1 \\ -5 & 8 \end{vmatrix} + \begin{vmatrix} 0 & -4 \\ 4 & -2 \\ 4 & -6 \end{vmatrix} = \begin{vmatrix} 4 & 4 \\ 3 & -3 \\ -1 & 2 \end{vmatrix}$$

47) Answer: C.

The area of 9 feet by 9 feet room is 81 square feet.

$9 \times 9 = 81$

"End"

www.ingramcontent.com/pod-product-compliance
Lightning Source LLC
Chambersburg PA
CBHW081326090426
42737CB00017B/3041